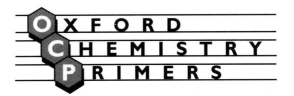

SERIES EDITOR
STEPHEN G. DAVIES
The Dyson Perrins Laboratory, University of Oxford

Oxford Chemistry Primers

1 S. E. Thomas *Organic synthesis: the roles of boron and silicon*
2 D. T. Davies *Aromatic heterocyclic chemistry*
3 P. R. Jenkins *Organometallic reagents in synthesis*
4 M. Sainsbury *Aromatic chemistry*
5 L. M. Harwood *Polar rearrangements*
6 I. E. Markó *Oxidations*

Aromatic Heterocyclic Chemistry

David T. Davies

SmithKline Beecham Pharmaceuticals, Harlow, Essex

OXFORD   NEW   YORK   TOKYO
OXFORD UNIVERSITY PRESS
1992

Oxford University Press, Walton Street, Oxford OX2 6DP

Oxford New York Toronto
Delhi Bombay Calcutta Madras Karachi
Petaling Jaya Singapore Hong Kong Tokyo
Nairobi Dar es Salaam Cape Town
Melbourne Auckland
and associated companies in
Berlin Ibadan

Oxford is a trademark of Oxford University Press

Published in the United Sates
by Oxford University Press, New York

A catalogue record for this book is available from the British Library

Library of Congress Cataloging-in-Publication Data

Davies, David T.
Aromatic heterocyclic chemistry/David T. Davies.
1. Heterocyclic chemistry.   I. Title.
QD400.D38   1991   547'.59—dc20   91-34831
ISBN 0-19-855661-6 (Hbk)
ISBN 0-19-855660-8 (Pbk)

Typeset by Pentacor Ltd, High Wycombe
Printed in Great Britain by
Information Press Ltd, Eynsham, Oxford, England

Series Editor's Foreword

Aromatic heterocyclic chemistry is an enormous and complex subject of great industrial and academic significance. A number of the molecules of life are derived from aromatic heterocycles and many important pharmaceutical and agrochemical compounds are based on aromatic heterocycles. Consequently, the importance of aromatic heterocyclic chemistry has stimulated a vast amount of synthetic and theoretical work in the area.

Oxford Chemistry Primers have been designed to provide concise introductions relevant to all students of chemistry, and contain only the essential material that would be covered in an 8–10 lecture course. In this primer David Davies has produced an excellent introduction to aromatic heterocyclic chemistry that should stimulate any reader to explore further into this vast topic. This primer will be of interest to apprentice and master chemist alike.

Stephen G. Davies
The Dyson Perrins Laboratory, University of Oxford

Preface

Heterocyclic chemistry is a vast discipline and at first sight impossible to do justice to in a text of this size. The aim of this book is to present only the essential features of the more important ring systems. Many reaction mechanisms are discussed in detail and several complete syntheses of heterocycles are presented. I hope that the reader will find this text both interesting and instructive, and that it will provide the platform for further study of this fascinating subject.

I would like to thank my friends and colleagues at SmithKline Beecham Pharmaceuticals for their helpful comments, including Angela Gadre, Clare Hayward, Chris Johnson, Helen Morgan and vacation students Peter Ainsworth and Francis Montgomery. I am similarly grateful to Prit Shah and colleagues of Glaxo Group Research. Finally I am indebted to Roger Martin of SmithKline Beecham Pharmaceuticals for helping me with the chemical structure drawing package.

Harlow
June 1991

D. T. D.

To Julie

Contents

1	Introduction	1
2	Pyrroles, thiophenes, and furans	10
3	Oxazoles, imidazoles, and thiazoles	20
4	Isoxazoles, pyrazoles, and isothiazoles	28
5	Pyridines	35
6	Quinolines and isoquinolines	46
7	Indoles	53
8	Five-membered ring heterocycles with three or four heteroatoms	61
9	Six-membered ring heterocycles containing one oxygen atom	67
10	Pyrimidines	73
11	Answers to problems	78
	Index	87

1. Introduction

1.1 Heterocyclic chemistry

Heterocyclic chemistry is a large and important branch of organic chemistry. Heterocycles occur in nature, for instance in nucleic acids (see Chapter 10) and indole alkaloids (see Chapter 7). Synthetic heterocycles have widespread uses as herbicides (e.g. **1.1**), fungicides (e.g. **1.2**), insecticides (e.g. **1.3**), dyes (e.g. **1.4**), organic conductors (e.g. **1.5**), and, of course, pharmaceutical products such as the anti-ulcer drug **1.6**.

1.2 Aromaticity and heteroaromaticity

Any ring system containing at least one heteroatom (i.e. an atom other than carbon – typically nitrogen, oxygen, or sulphur) can be described as heterocyclic. This broad definition encompasses both aromatic heterocycles (such as pyridine **5.1**) and their non-aromatic counterparts (piperidine **1.7**).

The compound numbering system in this chapter is not as odd as it might seem. For more on compound **5.1** see Chapter 5, etc.

Aromatic heterocycles are described as being heteroaromatic, and we shall concentrate on these systems in this book at the expense of more saturated systems. Let us now consider the concept of aromaticity with regard to benzene.

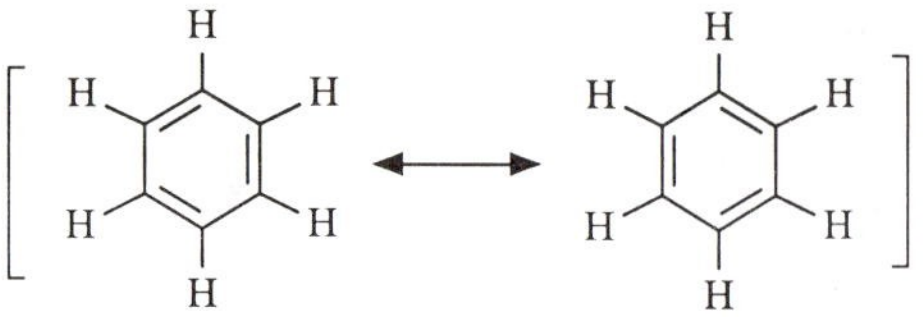

1.8a   **1.8b** 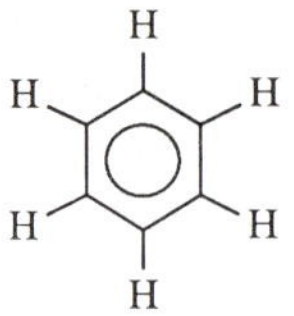   **1.9**

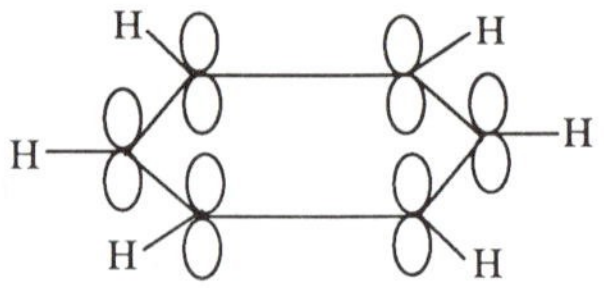

The carbon atoms in benzene are sp^2 hybridised, and the hydrogen atoms are in the same plane as the carbon atoms. The remaining six p orbitals are at right angles to the plane of the ring and contain six π electrons. Benzene fulfils the Hückel criteria for aromaticity as applied to cyclic polyenes containing 4n + 2 electrons (where $n=1$ in this case) in filled p orbitals capable of overlap.

Although two mesomeric representations **1.8a,b** can be drawn for benzene, this does *not* imply two rapidly-interconverting forms. Rather, the six π electrons are delocalised in molecular orbitals resulting in an annular electron cloud above and below the plane of the ring. Benzene can also be represented by structure **1.9**, which emphasises the cyclical arrangement of electrons. In agreement with this theory, the carbon–carbon bond lengths are all equivalent (0.14 nm) and intermediate between that of a single (0.154 nm) and double (0.133 nm) carbon–carbon bond. The extra thermodynamic stabilisation imparted to benzene by this phenomenon of electron delocalisation, called 'resonance', can be determined indirectly. Real, delocalised benzene is thermodynamically more stable than a theoretical cyclohexatriene molecule (i.e. non-delocalised structure **1.8a**) by around 150 kJ mol⁻¹.

How does this concept of aromaticity apply to typical heterocycles such as pyridine **5.1** and pyrrole **2.1**? Pyridine can formally be derived from benzene by replacement of a CH unit by an sp^2 hybridised nitrogen atom. Consequently, pyridine has a lone pair of electrons instead of a hydrogen atom. However the six π electrons are essentially unchanged, and the pyridine is a relatively aromatic heterocycle.

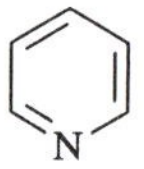

5.1    **2.1**

A difficulty arises with five-membered heterocycles such as pyrrole, which at first sight would appear to have only four π electrons, two short of the 4n + 2 Hückel criteria for aromaticity. The nitrogen atom is sp^2 hybridised and formally contains a lone pair of electrons in the remaining p orbital at right angles to the ring. However, the system is delocalised, as shown below.

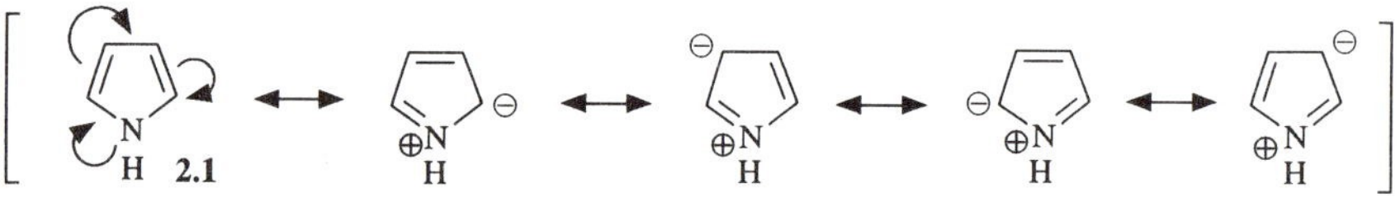

Thus, delocalisation of the nitrogen lone pair completes the sextet of electrons required for aromaticity. These two examples illustrate the point that certain heterocycles (closely analogous to benzene and naphthalene) such as pyridine **5.1**, pyrimidine **10.1,** and quinoline **6.1** are of aromatic 'by right' whereas other heterocycles such as pyrrole **2.1**, imidazole **3.2**, and triazole **8.7** have to 'earn' aromaticity by delocalisation of a lone pair of electrons from the heteroatom.

| 5.1 | 10.1 | 6.1 | 2.1 | 3.2 | 8.7 |

What are the consequences of this concept of lone pair delocalisation for a related series of heterocycles such as pyrrole **2.1**, thiophene **2.2**, and furan **2.3**? As delocalisation results in electron loss from the heteroatom concerned, the extent of delocalisation (and hence aromaticity) will vary with the electronegativity of the heteroatom. The highly electronegative oxygen atom in furan holds on to electron density more strongly than the heteroatom in thiophene or pyrrole. Furan is generally considered to have a non-aromatic electron distribution fairly close to that depicted by structure **2.3**.

| 2.1 | 2.2 | 2.3 |

In fact the thorny problem as to how aromatic is a particular heterocycle or series of heterocycles has been a preoccupation of physical organic chemists for some time. Bond lengths, heats of combustion, spectroscopic data, and theoretically-calculated resonance energies have all been invoked, but an absolute measure of aromaticity remains elusive. Nevertheless, trends regarding relative aromaticity will be alluded to in this text as they arise.

For a review on the concept of heterocyclic aromaticity see Katritzky *et al* (1991).

1.3 Synthesis of heterocycles

There are many syntheses of the major heterocycles and they are often complementary in that they afford different substitution patterns on the ring. Most of the synthetic methods we shall examine are fairly classical (indeed some are decidedly ancient!) although many of the specific examples are quite modern. Many classical syntheses of heterocycles revolve around the condensation reaction in its various guises. Let us consider the mechanism of a simple acid-catalysed condensation, that of generalised ketone **1.10** and amine **1.11** to give imine **1.12**.

Protonation of the ketone oxygen atom activates the ketone to nucleophilic attack by the amine. Loss of a proton from **1.13** produces neutral intermediate **1.14**. A second protonation, once again on the oxygen atom affords **1.15**, which on loss of a water molecule and a proton gives the imine **1.12**. All these steps are reversible, but in practice if water can be removed from the equilibrium (for instance by azeotropic distillation) then such reactions can be forced to completion. This type of reaction occurs many times in this text, but in future will not be presented in such detail. The student is strongly advised to work through, using pen and paper, the mechanism shown below and the many subsequent mechanisms. Confidence with reaction mechanisms will facilitate understanding of heterocyclic chemistry and organic chemistry in general.

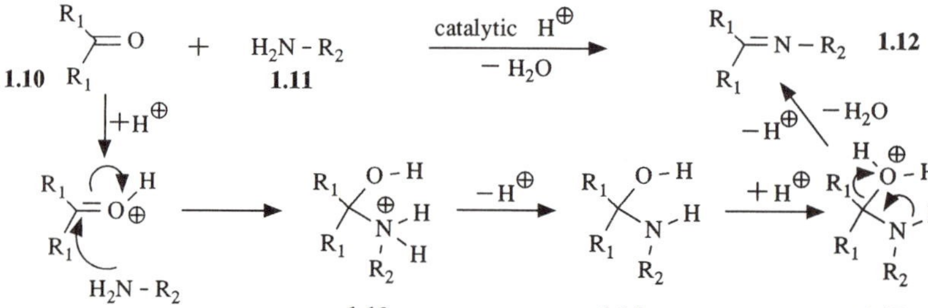

The disconnection approach to synthesis essentially involves working backwards from a target compound in a logical manner (so-called retrosynthesis), so that a number of possible routes and starting materials are suggested. This approach has been applied mainly to alicyclic, carbocyclic, and saturated heterocyclic systems. Retrosynthetic analyses are presented in this text not as an all-embracing answer to synthetic problems, but rather as an aid to understanding the actual construction of unsaturated heterocycles.

The symbol $\Rightarrow$ denotes a disconnection, an analytical process in which a structure is transformed into a suitable precursor

Returning to the condensation presented above, this leads to an important disconnection. The imine-like linkage present in several heterocycles (generalised structure **1.16**) can arise from cyclisation of **1.17**, containing amino and carbonyl functionalities.

Now consider condensation of ammonia with ketoester **1.18**. The isolated product is not imine **1.19** but the thermodynamically more stable enamine tautomer **1.20** which has a conjugated double bond system and a strong intramolecular hydrogen-bond. Although not a heterocyclic example, **1.20** illustrates that an enamine-like linkage, as in generalised heterocycle **1.21**, is also accessible by a condensation reaction.

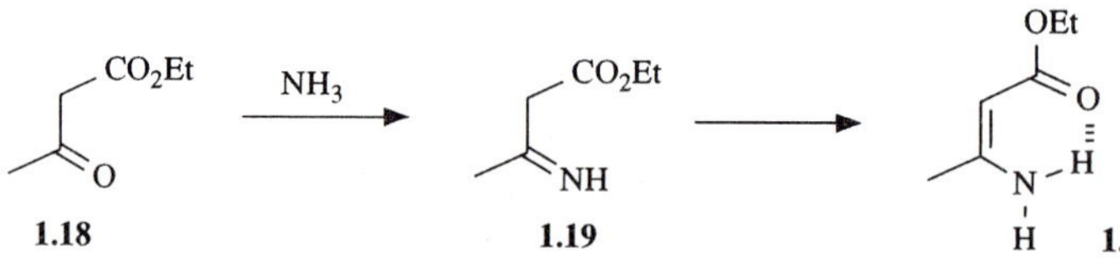

In a retrosynthetic sense, formal hydrolysis of the carbon-nitrogen bond of **1.21** reveals enol **1.22** which would exist as the more stable ketone tautomer **1.23**. Note that in the hydrolytic disconnection step the carbon becomes attached to a hydroxy group and the nitrogen to a hydrogen atom – there is no change in the oxidation levels of carbon or nitrogen.

Unlike our initial imine disconnection which is restricted to nitrogen heterocycles (with one or two specific exceptions such as pyrilium salts, see Chapter 9), the heteroatom in the enamine or enamine-like disconnection could be divalent. Therefore this disconnection is also applicable to oxygen- and sulphur-containing heterocycles, typified by **1.24** and **1.25**.

Let us see how this disconnection approach can rationalize the synthesis of pyrrole **2.16**.

Retrosynthetic analysis suggests a double condensation between diketone **1.26** and ammonia. Pyrrole **2.16** can actually be prepared if this way – see Chapter 2.2.

Another aid to understanding heterocyclic synthesis in general is the fact that a large number of five- and six-membered heterocycles can be constructed from various combinations of small acyclic molecules by complementary matching of nucleophilic and electrophilic functionality.

Returning to the synthesis of pyrrole **2.16**, diketone **1.26** can be regarded as a four-carbon bis-electrophilic fragment and ammonia, in this instance, as a bis-nucleophilic nitrogen fragment. Ammonia can form up to three bonds in a nucleophilic manner.

In this particular instance the correct oxidation level automatically results from the condensation reaction, giving pyrrole **2.16** directly. In other cases cyclisation does not afford the correct oxidation level and an unsaturated system has to be oxidised to achieve aromaticity. For instance, 1,5-diketones **1.27** react with ammonia to give dihydropyridines **1.28** which can be oxidised to pyridines **1.29**.

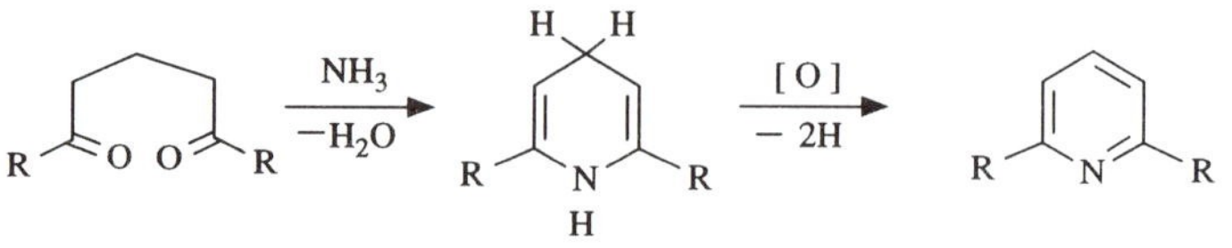

| **1.27** | **1.28** | **1.29** |

Examples of this cyclisation–oxidation strategy include the synthesis of pyridotriazine **5.32** (page 42) and syntheses of quinolines and isoquinolines (Chapter 6). Some examples of nucleophilic and electrophilic fragments are shown in Table 1.1. Several points arise from the table.

Consider acylating species such as acid chlorides. Acylation of diamine **1.30** initially gives amide **1.31** which undergoes a condensation to produce benzimidazole **1.32**. The carbonyl moiety is acting exclusively as an electrophilic centre.

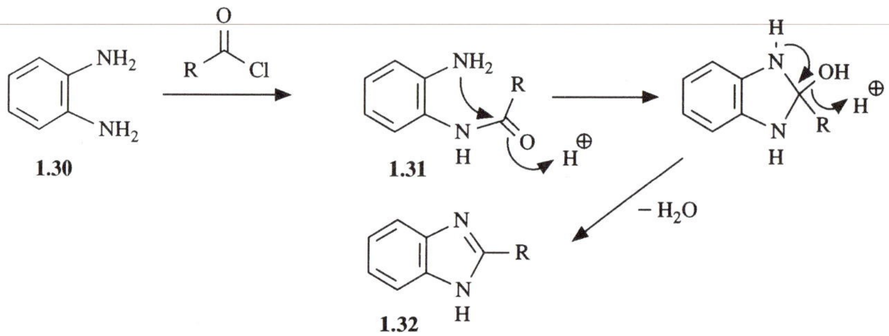

1.30   **1.31**   **1.32**

However, delocalisation of the nitrogen lone pair in the amide linkage (see mesomeric representations **1.33a,b**) produces a nucleophilic oxygen atom which can react with electrophiles as shown.

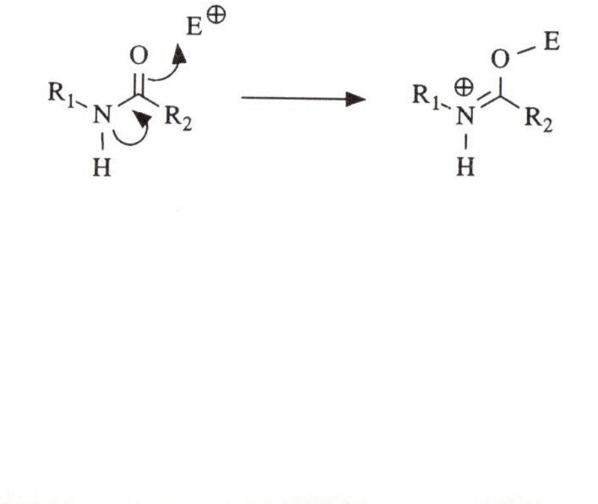

1.33a   **1.33b**

Table 1.1

Nucleophilic fragments
No. of ring atoms

1 NH_3 , H_2O , H_2S *(see Chapters 2 and 5)*

2 H_2N-NH_2 , H_2N-OH *(see pyrazole and isoxazole synthesis, Chapter 4)*

3

(see thiazole synthesis, Chapter 3, and pyrimidine synthesis, Chapter 10)

(see quinoline synthesis, Chapter 6)

4

(see benzimidazole synthesis, Chapter 1)

5

(see isoquinoline synthesis, Chapter 6)

Electrophilic fragments
No. of ring atoms

1

(X = leaving group, eg. Cl- see benzimidazole synthesis, Chapter 1 and isoquinoline synthesis, Chapter 6)

2

(see thiazole synthesis, Chapter 3)

3

(see quinoline synthesis, Chapter 6)

R_1, R_2 = alkyl or O-alkyl

(see pyrazole and isoxazole synthesis, Chapter 4, and pyrimidine synthesis, Chapter 10)

4

(see pyrrole, thiophene, and furan syntheses, Chapter 2)

Nucleophilic / Electrophilic fragments
No. of ring atoms

2

(see Chapter 1 and oxazole synthesis, Chapter 3)

(see Chapter 1 and coumarin synthesis, Chapter 9)

3

(see oxazole synthesis, Chapter 3, and Knorr pyrrole synthesis, Chapter 2)

Amides can cyclize in this manner as, for example, in the acylation of amino acids **1.34** to afford oxazolidinones **1.35**.

Acylating species are thus included in both electrophilic and nucleophilic/electrophilic categories in Table 1.1. For a related example see the synthesis of oxazoles in Chapter 3.

1,3-Dicarbonyl compounds, such as malonate derivatives, can also be classified under two categories. As well as reacting simply as a three-atom bis-electrophilic fragment (as in the synthesis of barbiturate **10.25** (page 77), an alternative reactivity is available. Condensation (by nucleophilic attack) of the active methylene carbon and electrophilic reaction at just one of the carbonyl groups is a two-atom nucleophilic/electrophilic profile, as seen in the preparation of coumarin **9.16**.

These concepts of retrosynthesis and heterocycle construction will help put the syntheses encountered in the following chapters into a broader perspective.

1.4 References

Textbooks

Acheson, R.M. (1967). *An introduction to the chemistry of heterocyclic compounds*, (2nd edn). Wiley, New York.

Paquette, L.A. (1966). *Principles of modern heterocyclic chemistry*. Benjamin, New York.

Joule, J.A. and Smith, G.F. (1979). *Heterocyclic chemistry*, (2nd edn). Van Nostrand Reinhold, New York.

Gilchrist, T.L. (1985). *Heterocyclic chemistry*. Longman, Harlow.

The first two (Acheson and Paquette) are still very good texts even today. Of the more recent pair, both are warmly recommended. Joule and Smith is possibly a more introductory text than Gilchrist, which contains many journal references and is pitched at the advanced undergraduate/postgraduate level. See Gilchrist for a discussion of the nucleophilic/electrophilic fragment approach to heterocyclic synthesis.

Warren, S. (1978). *Designing organic syntheses*, p.150–172. Wiley, Chichester.

Warren S. (1982). *Organic synthesis - the disconnection approach*, p. 3260–345. Wiley, Chichester.

Reference books and series

Coffey, S. (ed.) (1973 – 1986). *Heterocyclic compounds* (Vols. 4A – 4K of *Rodd's chemistry of carbon compounds*). Elsevier, Amsterdam.

Elderfield, R.C. (ed.) (1950 – 1967). *Heterocyclic chemistry*, Vols. 1 – 9. Wiley, New York.

Katritzky, A.R. and Boulton, A.J. (ed.) (1963 – 1989). *Advances in heterocyclic chemistry*, Vols. 1 – 45. Academic Press, Orlando.

Katritzky, A.R. and Rees, C.W. (ed.) (1984). *Comprehensive heterocyclic chemistry*, Vols. 1 – 8. Pergamon Press, Oxford.

Katritzky, A.R. *et al*, (1991). *Heterocycles*, **32**, 127–161.

Sammes, P.G. (ed.) (1979). *Heterocyclic chemistry* (Vol. 4 of *Comprehensive organic chemistry*, ed. D. Barton and W.D. Ollis). Pergamon Press, Oxford.

Weissburger, A. and Taylor, E.C. (ed.) (1950 – 1990). *The chemistry of heterocyclic compounds*. Wiley Interscience, New York.

All of these sources contain excellent reviews on virtually every aspect of heterocyclic chemistry. In particular, Katritzky and Rees is a thoroughly comprehensive work. For those particularly interested in the synthesis of heterocycles as pharmaceutical agents see:

Lednicer, D. and Mitscher, L.A. (1977, 1980, 1984, and 1990). *Organic chemistry of drug synthesis*, Vols. 1 – 4. Wiley, New York.

Experimental references

In this introductory text there is little detail regarding solvents, yields, workup procedures, etc. However, several chapters reference a relevant experimental procedure (taken from *Organic syntheses* or *Vogel*) which the student is strongly encouraged to read. For an excellent selection of experimental procedures for the synthesis of heterocycles see:

Furniss, B.S., Hannaford, A.J., Smith, P.W.G., and Tatchell, A.R. (1989). *Vogel's textbook of practical organic chemistry* (5th edn), pp. 1127 – 1194. Longman, Harlow.

2. Pyrroles, thiophenes, and furans

2.1 Introduction

The numbering of heterocycles generally starts at the heteroatom

Pyrrole **2.1**, thiophene **2.2**, and furan **2.3**, are five-membered ring heteroaromatic compounds containing one heteroatom. They derive their aromaticity from delocalisation of a lone pair of electrons from the heteroatom. Consequently the lone pair is not available for protonation and hence these heterocycles are not basic.

Under extreme conditions of acidity pyrrole is protonated, but at the C2 position.

2.1   **2.2**   **2.3**

The basis and extent of their aromaticity is discussed in Chapter 1. In summary, the capacity for the lone pair on a particular heteroatom to be delocalised is inversely related to the electronegativity of the heteroatom. For instance, furan is the least aromatic of the trio because oxygen has the greatest electronegativity and hence mesomeric representations **2.4b-e** make relatively less of a contribution to the electronic structure of furan than they do in the cases of pyrrole and thiophene. The order of aromaticity is furan < pyrrole < thiophene. We shall see later how this variation in aromaticity affects the reactivities of these three related heterocycles.

Note that protonation of the pyrrole nitrogen would lead to a non-aromatic cation.

2.4a   **2.4b**   **2.4c**   **2.4d**   **2.4e**

X = NH , S , O

A small number of simple pyrroles such as **2.5** and **2.6** occur naturally. Far more important are the tetramic pyrrole derivatives (porphyrins) such as chlorophyll-*a* **2.7** and haem **2.8**.

2.5   **2.6**

2.7

2.8

Chlorophyll-*a* is a plant pigment involved in the crucial photosynthetic process in which the energy of sunlight is harnessed to incorporate carbon dioxide into plant metabolism. Haem, however, is fundamental to mammalian biology, being the oxygen-binding component of haemoglobin. Oxygen absorbed from the air is transported around the body while temporarily co-ordinated to the iron atom of haemoglobin, which occurs in the red blood cells.

Acetylenic thiophene **2.9**, found in some species of higher plants, is one of the few naturally-occurring thiophenes. However, the thiophene ring is used in several important pharmaceutical products, such as the penicillin antibiotic **2.10**.

2.9

2.10

In contrast to the pyrrole and thiophene series, the furan nucleus occurs in many plant-derived terpenes such as **2.11**. The most important furan-containing drug is **2.12**, which reduces gastric acid secretion and is important in the treatment of ulcers.

Terpenes are plant-derived natural products constructed of multiples of the five-carbon hydrocarbon isoprene.

2.11

2.12

2.2. Synthesis of pyrroles, thiophenes, and furans

We shall first examine a general synthesis applicable to all three heterocycles, then consider two specific syntheses of pyrroles.

Retrosynthetic cleavage of a carbon–heteroatom bond in **2.13** gives enol **2.14** which is equivalent to ketone **2.15**. Repeating the process gives us a 1,4-dicarbonyl compound and the heteroatom-containing fragment such as a primary amine or hydrogen sulphide.

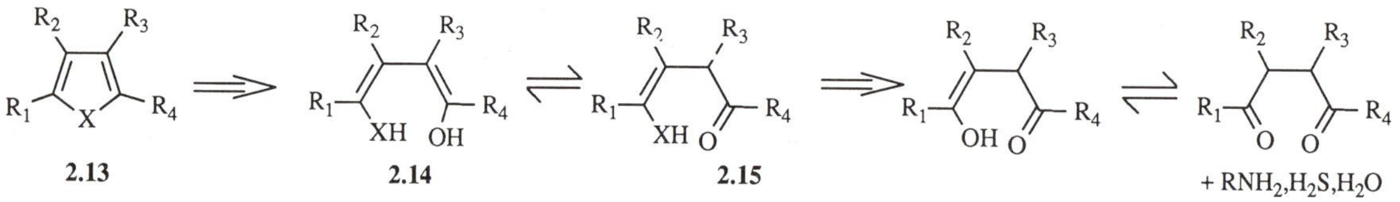

2.13   **2.14**   **2.15**   + RNH$_2$,H$_2$S,H$_2$O

The forward process is known as the Paal–Knorr synthesis. This is a very straightforward synthesis limited only by the accessibility of the 1,4-dicarbonyl precursors.

The mechanism is illustrated by the preparation of 2,5-dimethyl pyrrole **2.16** and is simply two consecutive condensations.

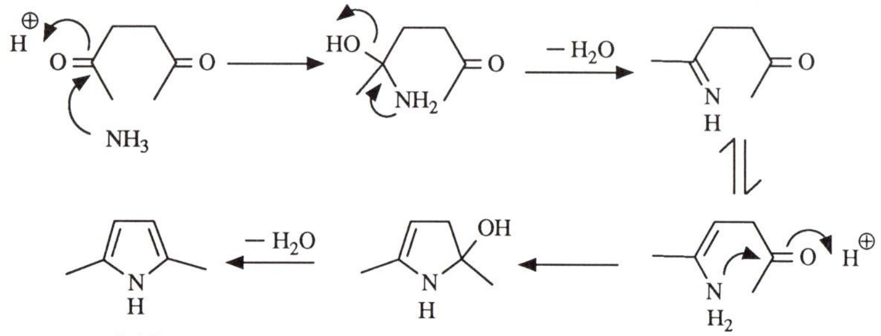

2.16

The Paal–Knorr synthesis can similarly be applied to thiophenes, e.g. compounds **2.17** - **2.20**.

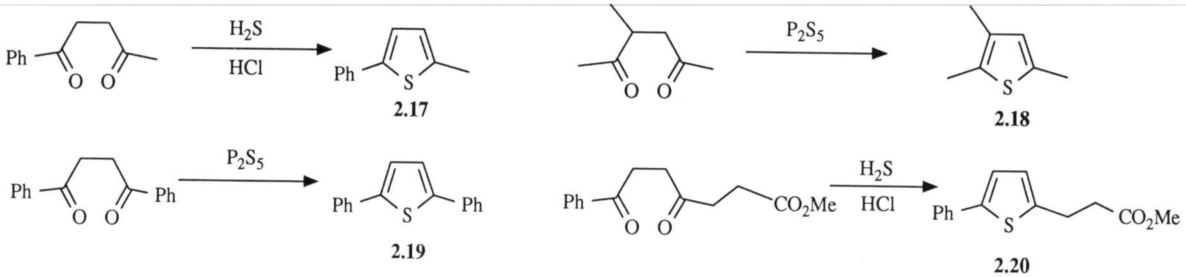

2.17

2.18

2.19

2.20

When hydrogen sulphide is the heteroatom source the mechanism is similar to the pyrrole case.

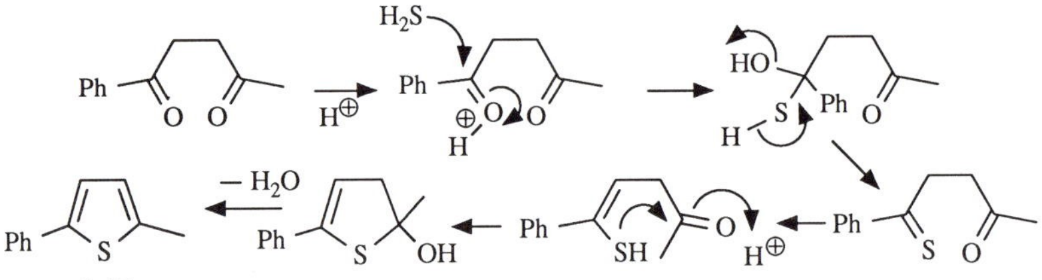

2.17

However, the situation is slightly different when phosphorous (V) sulphide is used. This reagent converts ketones to thioketones, by exchange of a phosphorus–sulphur double bond with a carbon–oxygen double bond.

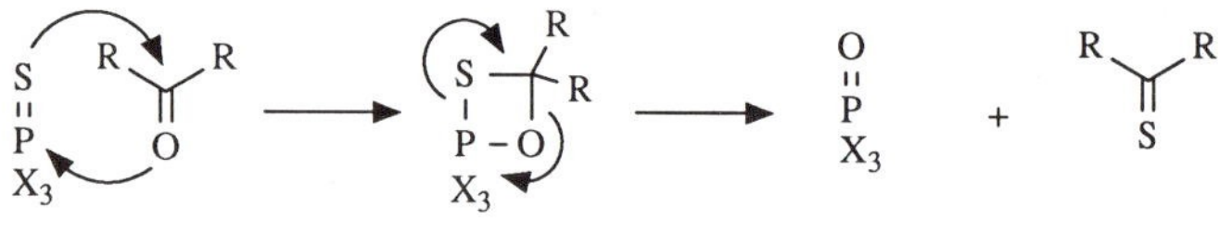

For instance, in the synthesis of **2.19**, the 1,4-diketone is converted into the corresponding 1,4-dithioketone followed by loss of hydrogen sulphide.

The mechanism of the cyclisation step is similar to that of thiophene **2.17**.

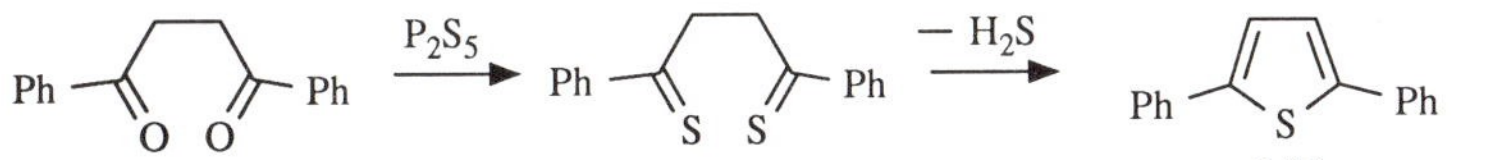

Our retrosynthetic analysis of the Paal–Knorr synthesis leads to a problem when applied to furan, as it implies addition of a water molecule, followed by elimination of two water molecules. In practice, simple dehydration of a 1,4-dicarbonyl compound leads to furans as in the preparation of **2.21**.

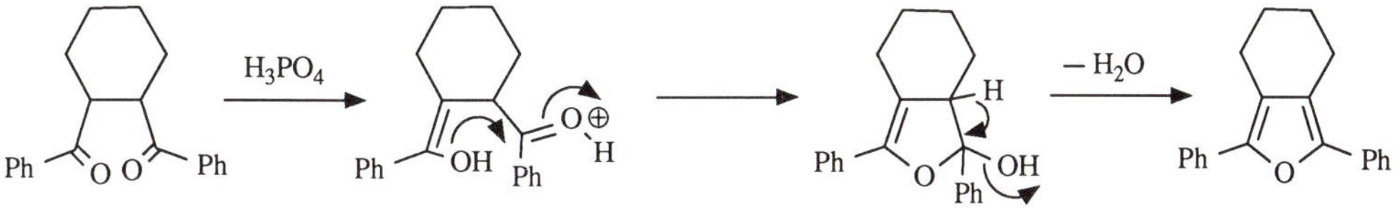

Returning again to pyrroles, probably the most widely-used method for their preparation is the Knorr pyrrole synthesis, which is the condensation of a ketone **2.22** with an α-aminoketone **2.23** to give pyrrole **2.13**, via enamine **2.24**. A reasonable mechanism is shown below, although none of the intermediates is isolated.

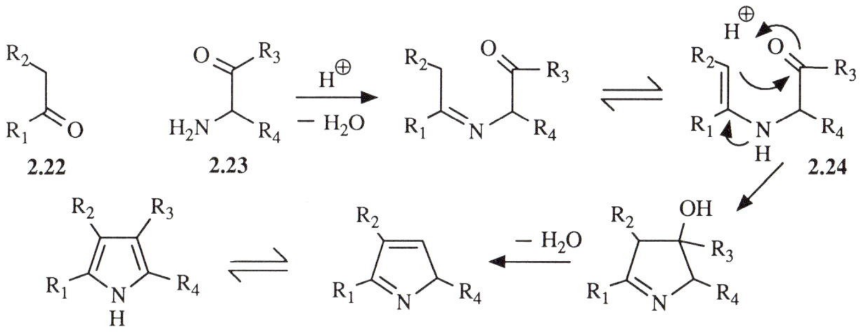

The α-aminoketones are often prepared by nitrosation of an active methylene group followed by reduction of the oxime to the amine (e.g. **2.25** to **2.26** to **2.27**).

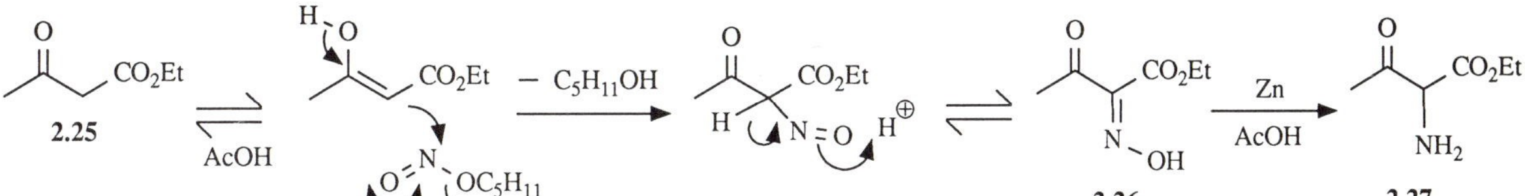

As α-aminoketones are prone to self-condensation (see page 22 for a discussion of α-aminoketones), the initial condensation step is facilitated by R_2 in **2.22** being an electron-withdrawing group. This enhances the electrophilic nature of the ketone carbonyl group thereby increasing the rate of the desired reaction, and favours enamine tautomer **2.24** over the imino tautomer because of conjugation with the electron-withdrawing group. A selection of Knorr pyrrole syntheses, showing the key intermediate enamines, is shown below.

The Knorr pyrrole synthesis consists of a ketone and amine condensing to give an enamine, followed by intramolecular cyclisation of this enamine onto the remaining ketone.

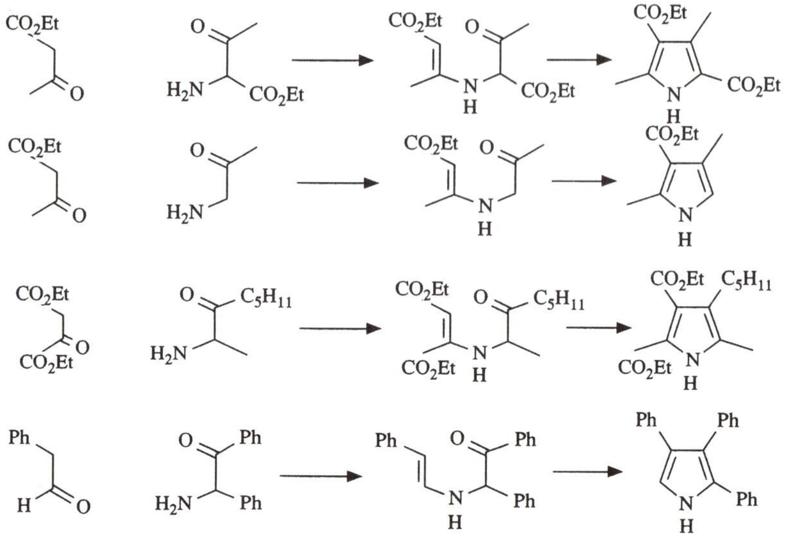

2.3 Electrophilic substitution of pyrrole, thiophene, and furan

Note that pyrrole reacts with electrophiles on carbon, like an enamine.

All three heterocycles undergo aromatic substitution reactions, though their reactivities vary considerably. Let us consider a generalised mechanism and how the stability of the two possible intermediates affects the position of substitution.

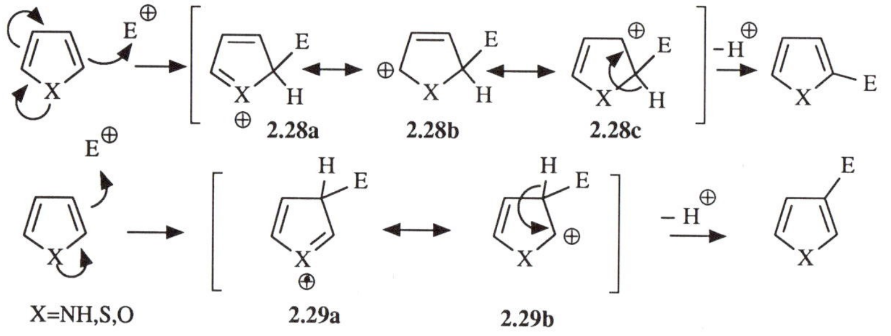

The intermediate derived from attack at the C2 position has greater delocalisation of the positive charge (mesomeric forms **2.28a,b,c**) than that derived from attack at the C3 position (mesomeric forms **2.29a,b**). As the charge is more extensively delocalised in the former, this intermediate is at lower energy. This in turn is reflected in a lower activation energy for this pathway and manifested in a selectivity for electrophilic substitution at the C2 position over the C3 position. The actual isomer ratio depends on the heterocycle, the electrophile, and the precise conditions, although in many cases such reactions are virtually regiospecific, and only the C2 substitution products are isolated. Very reactive electrophiles (such as the nitronium ion NO_2^+) exhibit lower selectivity because they tend to be less discriminating as to where they attack the heteroaromatic nucleus.

The ease of electrophilic substitution is pyrrole > furan > thiophene > benzene.

Pyrrole is extremely reactive towards electrophiles while thiophene, the most aromatic of the trio, is much less reactive. At a very rough approximation, the reactivity of thiophene is of the order of a heteroatom-substituted benzene derivative such as phenol. Despite large differences in the rates of electrophilic substitutions there are some important aromatic substitution reactions common to all three heterocycles.

The Vilsmeier reaction is the formylation of reactive aromatic compounds by using a combination of phosphorous oxychloride and N,N-dimethylformamide, followed by a hydrolytic workup.

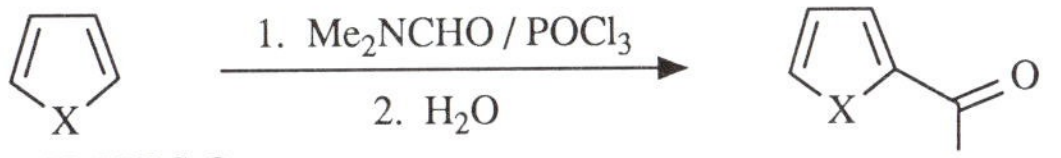

To give a quantitative feel for these differences in reactivity, data for the bromination of three representative derivatives are shown below.

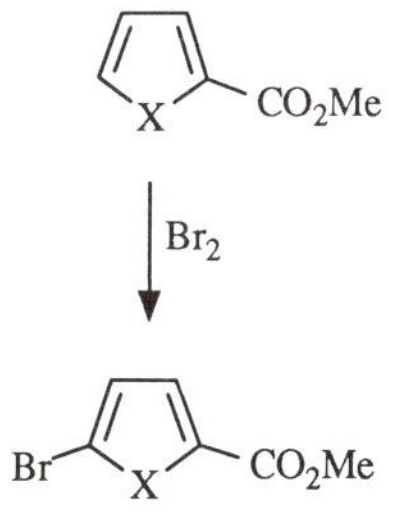

X	Relative Rate
S	1
O	1.2×10^2
NH	5.6×10^8

The reaction proceeds by formation of the electrophilic Vilsmeier complex **2.30**, followed by electrophilic substitution of the heterocycle. The formyl group is generated in the hydrolytic workup. Pyrrole, thiophene, and furan all undergo this formylation which is highly selective for the C2 position.

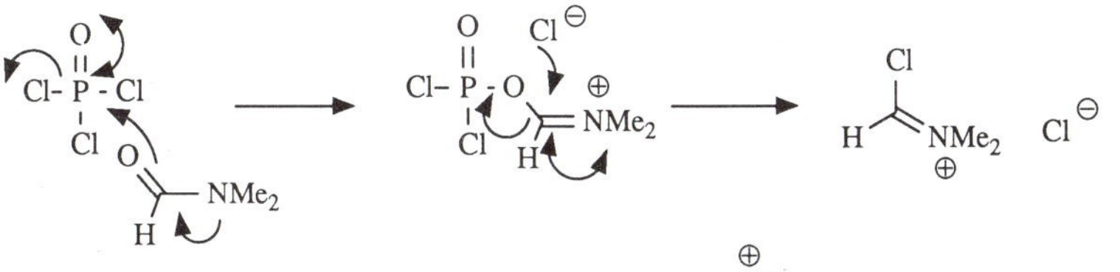

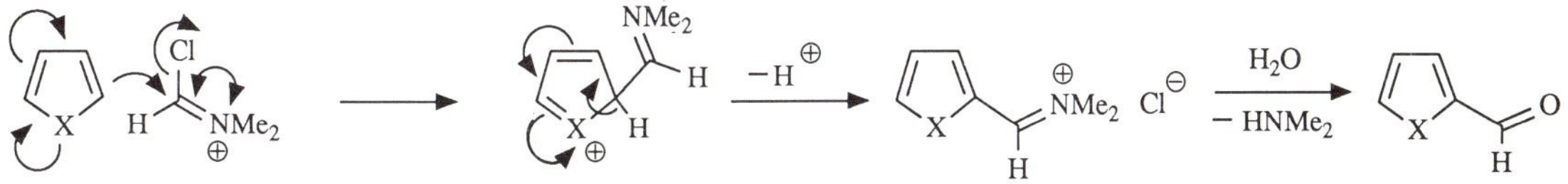

All three heterocycles undergo sulphonation with the pyridine–sulphur trioxide complex. This behaves like a mild source of sulphur trioxide, enabling the sulphonation to be carried out under essentially neutral conditions.

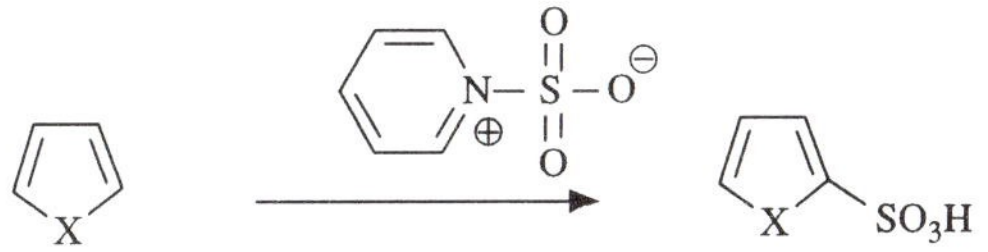

Furan and pyrrole are not stable to mineral acids, but acetyl nitrate can be used for the nitration of all three heterocycles.

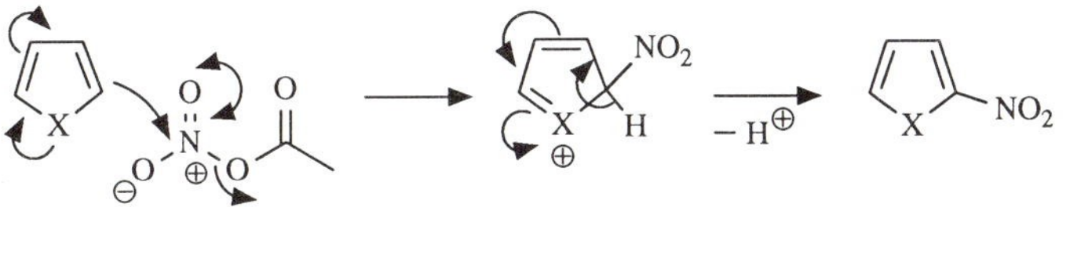

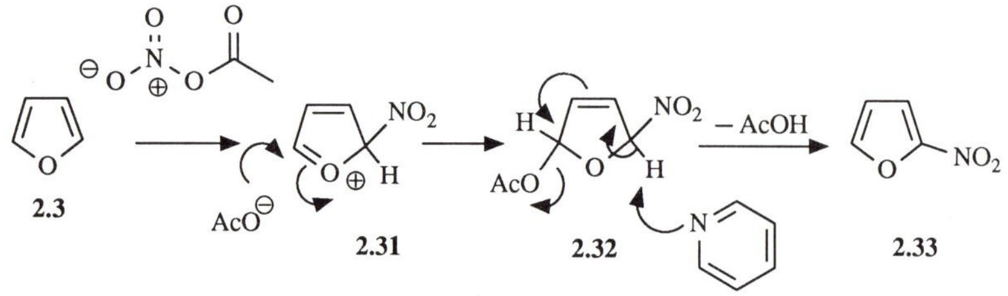

Whilst the mechanism shown above applies to pyrrole and thiophene, the nitration of furan with acetyl nitrate gives the 2,5-addition product **2.32**, arising from attack of acetate ion on the intermediate cation **2.31**. Treatment of **2.32** with pyridine eliminates the elements of acetic acid producing nitrofuran **2.33**.

The formation of **2.32** raises an important theoretical point: because furan is not very aromatic and the driving force to 're-aromatise' by loss of a proton is not very strong, cation **2.31** can be intercepted to give **2.32.** This behaviour is not observed with pyrrole and thiophene.

2.3 **2.31** **2.32** **2.33**

Thiophene, alkyl-substituted furans, and especially pyrrole, undergo Mannich reactions.

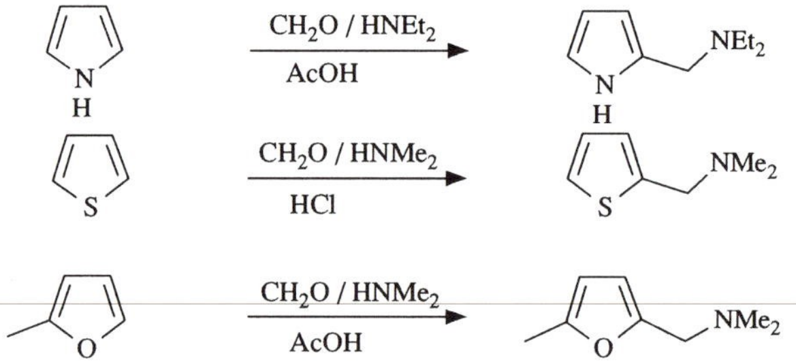

This involves condensation of the heterocycle, formaldehyde, and an amine (usually a secondary amine) to give an aminomethyl derivative.

$$Me_2NH + CH_2O \xrightarrow[-\,H_2O]{AcOH} CH_2{=}\overset{\oplus}{N}Me_2 \quad {}^{\ominus}OAc$$

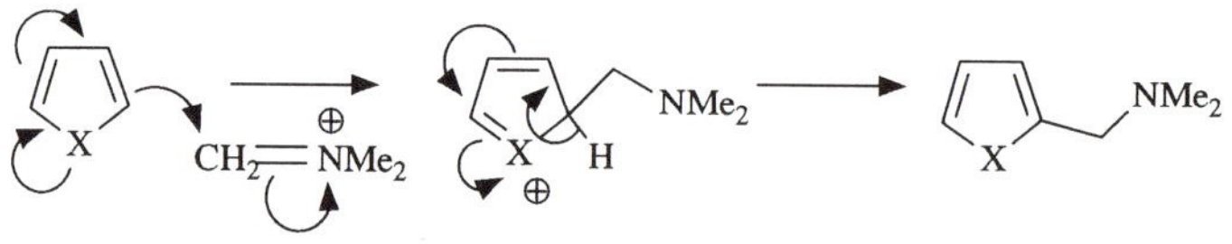

The Friedel–Crafts acylation and alkylation reactions are fundamental processes in aromatic chemistry. Pyrroles and furans are not stable to the Lewis acids necessary for these reactions, but thiophenes are stable to Lewis acids, and do undergo Friedel–Crafts acylation and alkylation.

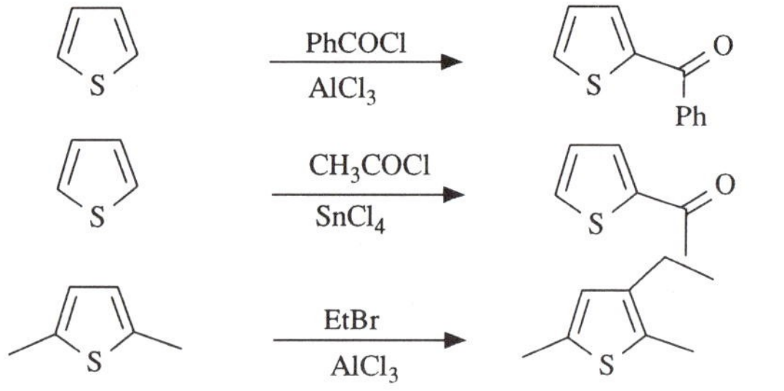

Observe that electrophilic substitution occurs at the C3 position when both the C2 and C5 positions are blocked.

The reactivity of all three heterocyles is considerably reduced when electron-withdrawing groups are present on the ring. This is important in the synthesis of pyrrole derivatives as it adds chemical stability to the ring, enabling reactions to be performed in the presence of Lewis acids.

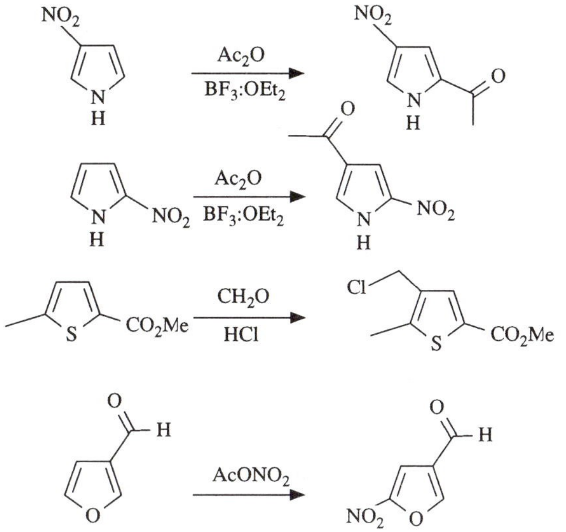

The regiochemistry of these reactions is easily explained by rationalisations from classical benzene chemistry, i.e. electron-withdrawing groups direct *meta*.

2.4 Anion chemistry of pyrroles, thiophenes, and furans

Pyrrole has a weakly acidic hydrogen atom attached to the nitrogen (pK_a= 17.5) and can be deprotonated by strong bases. The sodium and potassium salts are ionic in character and the naked anion tends to react on nitrogen as in the preparation of N-methyl pyrrole **2.34**. The corresponding magnesium derivative **2.35** has more covalent character and tends to react more on carbon than nitrogen, as in the preparation of pyrrole aldehyde **2.36**

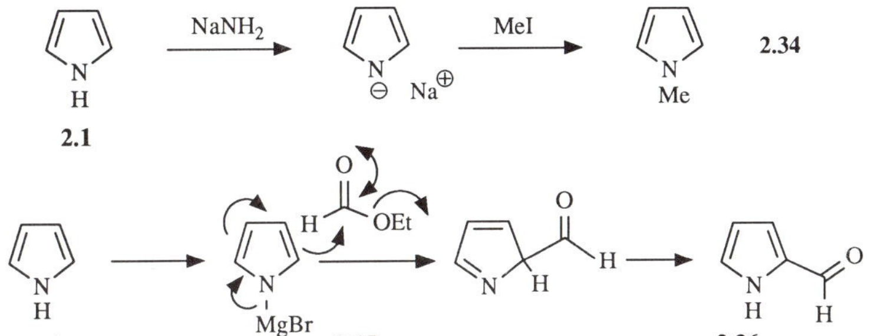

N-methyl pyrrole **2.34**, thiophene, and furan can be metallated at the C2 position with alkyl lithium reagents. This position is more activated to deprotonation than the C3 position because of the electron-withdrawing inductive effect of the heteroatom. The nucleophilic 2-lithio species can then be reacted with various electrophiles, as in the preparation of **2.37**, **2.38**, and **2.39**. Let us see how this methodology can be applied to the synthesis of **2.42**, a furan-containing mimic of a long-chain fatty acid. Deprotonation of furan and alkylation produces **2.39**. A second deprotonation at the C5 position and alkylation gives bromide **2.40**. Displacement of the bromide affords nitrile **2.41**, and acidic hydrolysis yields the target furan **2.42**.

The precise nature of the carbon–lithium bond is beyond the scope of this book. Organolithium intermediates are here represented as carbanion and cation to emphasise differences in properties and reactivities as compared with full covalent bonds.

The alkyl group at the C2 position is not deprotonated in the second alkylation.

Note the use of ⁻CN as a synthon for ⁻CO₂H.

2.5 Problems

1.

2.43

Tricyclic pyrrole derivative **2.43** is a drug currently under development for the treatment of schizophrenia. It is prepared by a Knorr pyrrole synthesis. What are the structures of the two starting materials required, and that of the intermediate enamine?

2. Why is pyrrole aldehyde **2.44** less reactive to nucleophiles than, say, benzaldehyde? Why is pyrrole alcohol **2.45** readily polymerised on exposure to acid?

2.44   **2.45**

3. Nitration of furan with nitronium tetrafluoroborate produces nitrofuran **2.33** directly. Contrast this result to the two stage reaction necessary with acetyl nitrate, page 16. Explain these observations.

4. What is the mechanism of this reaction?

2.6 References

Dean, F.M. (1982). *Adv. heterocyclic chem.*, **30**, 167; **31**, 237 (furans).

Gronowitz, S. (ed.) (1985). In *Thiophene and its derivatives (The chemistry of heterocyclic compounds* [ed. A. Weissburger and E.C. Taylor], Vol. 44). Wiley Interscience, New York.

Furniss, B.S., Hannaford, A.J., Smith, P.W.G, and Tatchell, A.R. (1989). *Vogel's textbook of practical organic chemistry* (5th edn), p.1148 (preparation of pyrrole **2.16**). Longman, Harlow.

Jackson, A.H. (1979). In *Heterocyclic chemistry* (ed. P.G. Sammes) (Vol. 4 of *Comprehensive organic chemistry*, ed. D. Barton and W.D. Ollis) (pyrroles). Pergamon Press, Oxford.

Jones, R.A. (ed.) (1990). In *Pyrroles (The chemistry of heterocyclic compounds* [ed. A. Weissburger and E.C. Taylor], Vol 48, Part 1). Wiley Interscience, New York.

Jones, R.A. and Bean, G.P. (1977). *The chemistry of pyrroles.* Academic Press, London.

Jones, E. and Moodie, I.M. (1970). *Org. syn.*, **50**, 104 (C2 metallation of thiophene).

Katritzky, A.R. and Rees, C.W. (ed.) (1984). *Comprehensive heterocyclic chemistry,* Vol. 4, Part 3 (five-membered rings with one heteroatom). Pergamon Press, Oxford.

Meth-Cohn, O. (1979). In *Heterocyclic chemistry* (ed. P.G. Sammes) (Vol. 4 of *Comprehensive organic chemistry*, ed. D. Barton and W.D. Ollis), p.737 (thiophenes). Pergamon Press, Oxford.

Sargent, M.V. (1979). In *Heterocyclic chemistry* (ed. P.G. Sammes) (Vol. 4 of *Comprehensive organic chemistry*, ed. D. Barton and W.D. Ollis), p.693 (furans). Pergamon Press, Oxford.

Silverstein, R.M., Ryskiewicz, E.E., and Willard, C. (1963). *Organic syntheses,* Coll. Vol. IV, 831 (Vilsmeier formylation of pyrrole).

3. Oxazoles, imidazoles, and thiazoles

3.1 Introduction

Oxazole **3.1**, imidazole **3.2**, and thiazole **3.3** are the parent structures of a related series of 1,3-azoles containing a nitrogen atom plus a second heteroatom in a five-membered ring.

3.1 **3.2** **3.3**

They are isomeric with the 1,2-azoles isoxazole, pyrazole, and isothiazole (see Chapter 4). Their aromaticity derives from delocalisation of a lone pair from the second heteroatom, **3.4a–e**.

3.4a **3.4b** **3.4c** **3.4d** **3.4e**

X=O,NH,S

The biosynthesis of histamine involves decarboxylation of the amino acid histidine.

The imidazole ring occurs naturally in histamine **3.5**, an important mediator of inflammation and gastric acid secretion. A quaternised thiazole ring is found in the essential vitamin thiamin **3.6**. There are few naturally occurring oxazoles, apart from some secondary metabolites from plant and fungal sources.

3.5 **3.6**

Oxazole, imidazole, and thiazole can be formally derived from furan, pyrrole, and thiophene respectively by replacement of a CH group by a nitrogen atom at the 3 position. The presence of this pyridine-like nitrogen deactivates the 1,3-azoles towards electrophilic attack and increases their susceptibility towards nucleophilic attack (see later). These 1,3-azoles can be viewed as hybrids between furan, pyrrole, or thiophene, and pyridine.

Imidazole (pK_a=7.0) is more basic than oxazole (pK_a=0.8) or thiazole (pK_a=2.5). This increased basicity arises from the greater electron-releasing capacity of two nitrogen atoms relative to a combination of nitrogen and a heteroatom of higher electronegativity. Also note that a symmetrical resonance-stabilised cation **3.7a,b** is formed.

The statement that oxazole has a pK_a of 0.8 means that the **protonated form** of oxazole is a very strong acid.
Therefore oxazole (as the free base) is a very weak base indeed.

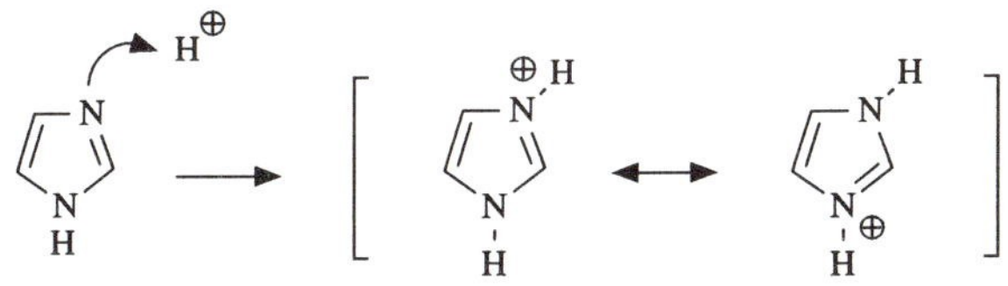

3.7a   **3.7b**

Furthermore, certain substituted imidazoles can exist in two tautomeric forms.

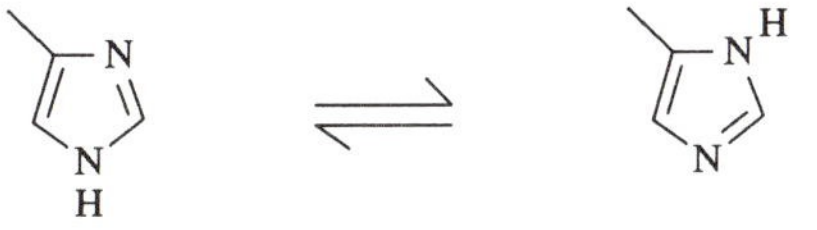

3.8   **3.9**

For instance, the imidazole shown above exists as a rapidly equilibrating mixture of 4-methyl **3.8** and 5-methyl **3.9** tautomers, and is referred to as 4(5)-methylimidazole. It must again be stressed that tautomerisation and resonance are totally different. Mesomeric representations **3.7a,b** are not interconverting like tautomers **3.8** and **3.9**; this is simply a means to describe an intermediate hybrid structure.

3.2 Synthesis of oxazoles

Retrosynthetic cleavage of the carbon–oxygen bond in generalised oxazole **3.10** produces iminoalcohol **3.11** (better represented in the amide form

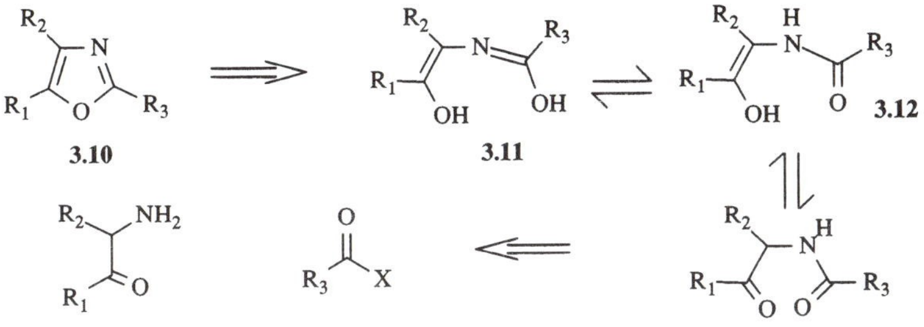

3.10   **3.11**

3.15   **3.14**

3.13

3.12). Similar tautomerisation of the enol group gives an actual intermediate **3.13**, and disconnection of the amide linkage reveals aminoketone **3.15** and an acylating species **3.14** such as an acid chloride. The forward process, cyclocondensation of amides **3.13** to yield oxazoles **3.10,** is known as the Robinson–Gabriel synthesis.

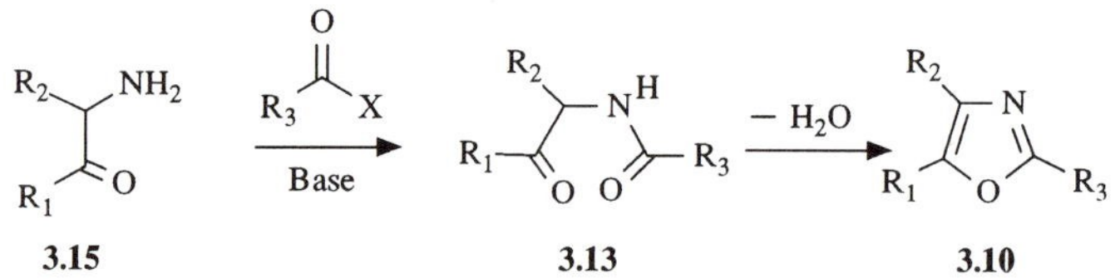

Base
$- H_2O$
3.15
3.13
3.10

In practice the dehydration can be achieved with a broad range of acids or acid anhydrides, such as phosphoric acid, phosphorous oxychloride, phosgene ($COCl_2$), and thionyl chloride. An example of the mechanism is shown below for thionyl chloride and involves activation of the amide to imidolyl halide **3.16** then intramolecular attack by the enolic form of the ketone.

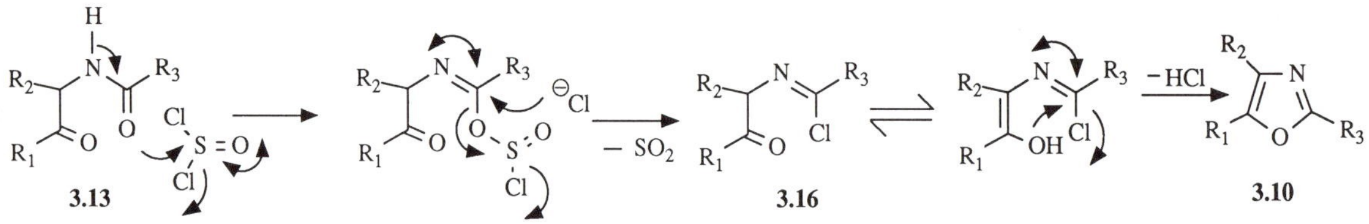

$- SO_2$
$- HCl$
3.13
3.16
3.10

The aminoketones thernselves can be prepared by a number of methods. A typical route is illustrated by the synthesis of anti-inflammatory drug **3.23**.

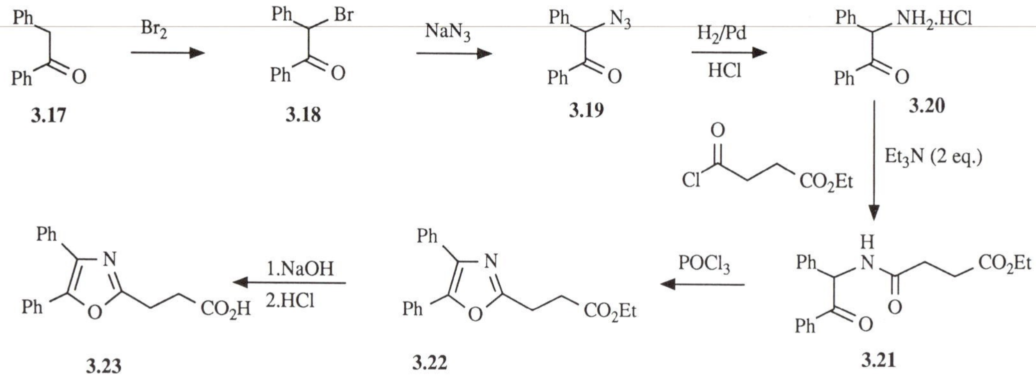

Br_2
NaN_3
H_2/Pd
HCl
Et_3N (2 eq.)
$POCl_3$
1.NaOH
2.HCl
3.17
3.18
3.19
3.20
3.21
3.22
3.23

Drugs which reduce inflammation are often used to treat the symptoms of arthritis.

Bromination of ketone **3.17** gives **3.18** which can be converted to azide **3.19**. Hydrogenation of **3.19** in the presence of hydrochloric acid affords aminoketone hydrochloride salt **3.20**. Such aminoketones are often isolated as the corresponding salts because the free aminoketones are prone to dimerisation, having both nucleophilic and electrophilic centres. (For a common alternative preparation of aminoketones, see the Knorr pyrrole synthesis, Chapter 2.) Liberation of the free base of **3.20** in the presence of the acid chloride affords amide **3.21** which is cyclised to oxazole **3.22**. Ester hydrolysis then affords the biologically-active carboxylic acid **3.23**.

3.3 Synthesis of imidazoles

Although there are several ways of preparing imidazoles, there is no one outstanding method. One useful synthesis is the condensation of a 1,2-

dicarbonyl compound with ammonium acetate and an aldehyde, as in the preparation of imidazole **3.25**.

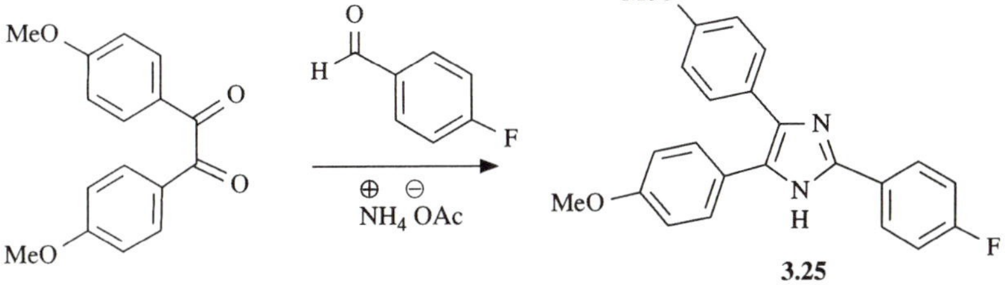

3.25

A reasonable rationalisation is a cyclocondensation type of process to give **3.24** followed by irreversible tautomerisation to **3.25**.

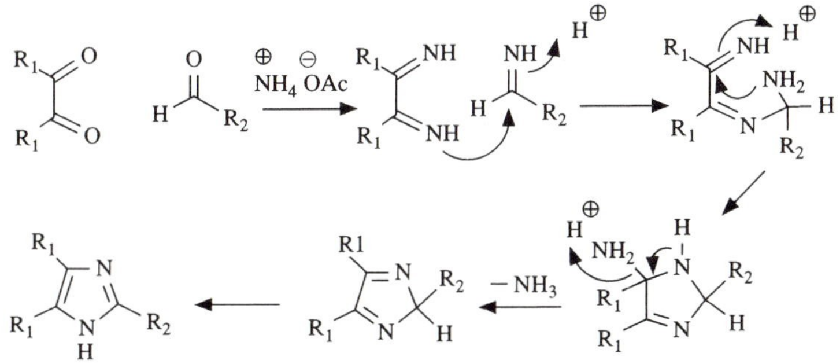

3.25   **3.24**

3.4 Synthesis of thiazoles

Retrosynthetic disconnection of the nitrogen–carbon bond in thiazole **3.26** leads formally to enol **3.27** which is equivalent to ketone **3.28**. This can be derived from haloketone **3.29** and thioamide **3.30**.

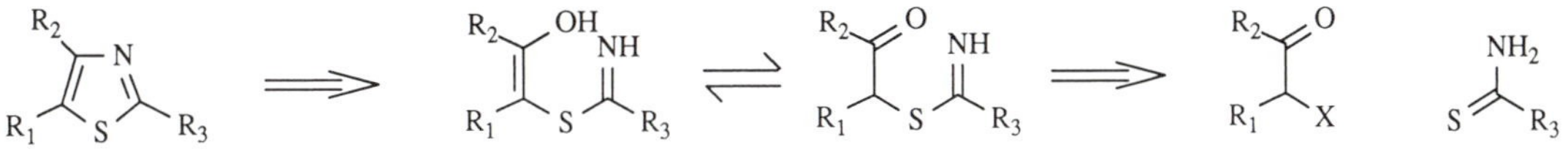

3.26   **3.27**   **3.28**   **3.29**   **3.30**

X=Cl,Br,I

The forward process is the Hantzsch synthesis of thiazoles which, despite its antiquity (it is around 100 years old), is still very widely used.

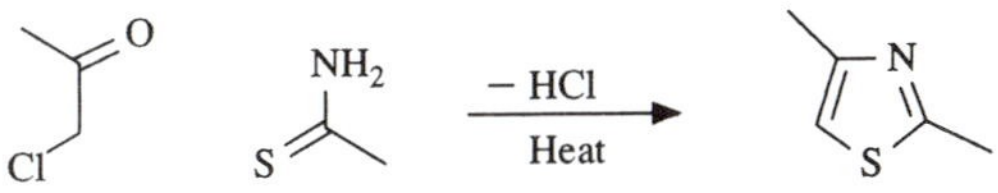

3.31

The mechanism for the formation of thiazole **3.31** involves initial nucleophilic attack by sulphur followed by a cyclocondensation.

Thiocarbonyl compounds are much more nucleophilic than carbonyl compounds.

3.31

The thioamides themselves are conveniently prepared from the corresponding amides by treatment with phosphorous (V) sulphide (see the Paal–Knorr synthesis of thiophenes, Chapter 2, for this type of conversion). A variation of the Hantzsch reaction utilises thioureas, where R_3 in **3.30** is a nitrogen and not a carbon substituent. For instance, thiourea itself is used in the preparation of 2-aminothiazoles such as **3.32**.

3.32

3.5 Electrophilic substitution reactions of oxazoles, imidazoles, and thiazoles

As with pyridine, not only does the electronegative nitrogen atom withdraw electron density from the ring, but under the acidic conditions of many electrophilic reactions the azole nitrogen is protonated. The azolium cation is relatively inert to further attack by a positively charged electrophile.

The 1,3-azoles are not very reactive towards electrophilic attack due to the deactivating effect of the pyridine-like nitrogen. However, electron-donating groups can facilitate electrophilic attack, as in the preparation of oxazoles **3.34** and **3.35**. Dimethylamino oxazole **3.33** is essentially functioning like an enamine in this reaction.

3.33   **3.34**

3.35

Imidazole can be nitrated under forcing conditions, nitration remarkably occurring on the imidazolium cation **3.7a,b**, giving nitroimidazole **3.36** after loss of two protons.

3.7ab   **3.36**

3.6 Anion chemistry of oxazoles, imidazoles, and thiazoles

The C2 position of 1,3-azoles is particularly electron-deficient because of the electron-withdrawing effect of the adjacent heteroatoms. The acidity of the protons at this position is such that deprotonation can be achieved with strong bases to give nucleophilic carbanions **3.37** which can be quenched with electrophiles producing substituted 1,3-azoles **3.38**.

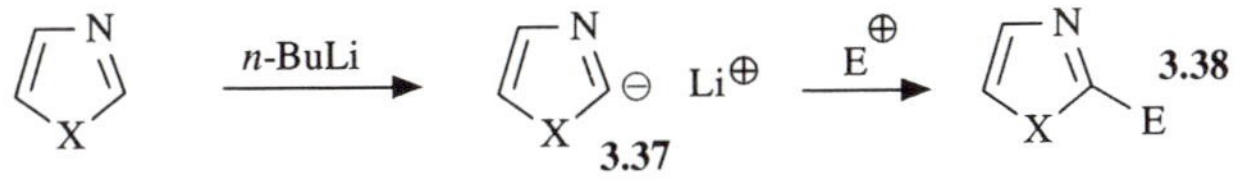

Similarly, alkyl groups at the C2 positions (but not the C4 or C5 positions) can be deprotonated giving carbanions **3.39a,b** which can also be quenched with electrophiles to afford 1,3-azoles **3.40**.

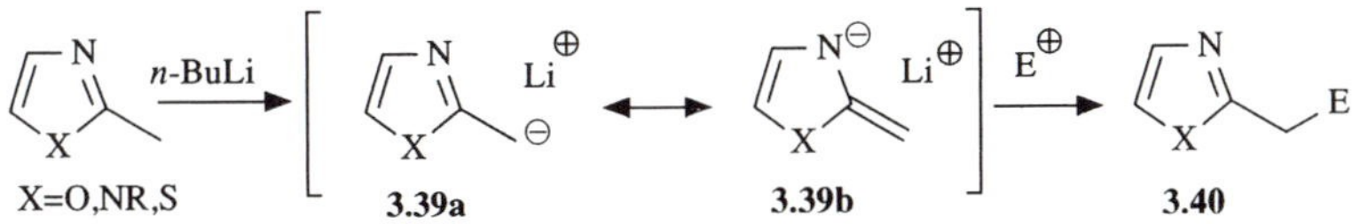

There is a useful analogy between resonance-stabilised anion **3.39a,b** and an ester enolate anion. Note that in both cases the negative charge can be delocalised onto a heteroatom.

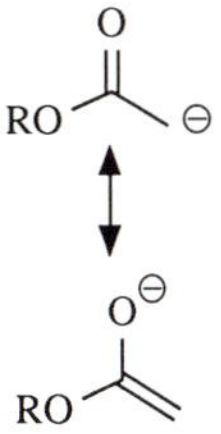

Some examples of both the above types of reactivity are given below

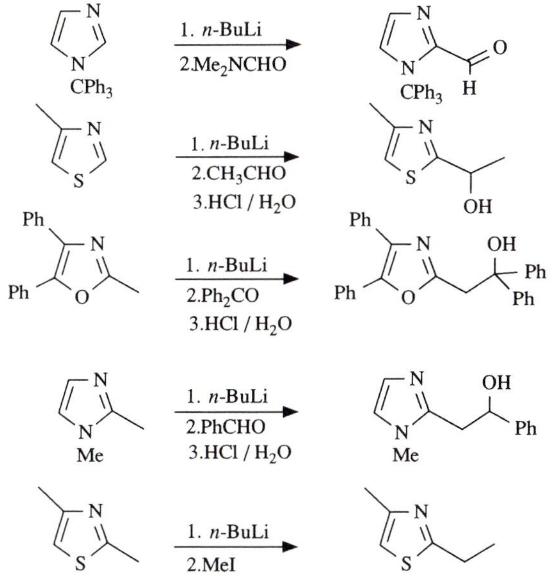

3.7 Nucleophilic aromatic substitution of oxazoles, imidazoles, and thiazoles

We have previously discussed the reduced reactivity to electrophiles of oxazole, imidazole, and thiazole, as compared to furan, pyrrole, and thiophene, which results from the presence of the pyridine-like nitrogen atom. This behaviour is paralleled by increased reactivity to nucleophiles. Nucleophilic attack on furan, pyrrole, and thiophene derivatives only occurs when an additional activating group is present, as in the displacement reaction giving thiophene **3.41**.

3.41

The nitro group plays a key role as an electron-acceptor in this reaction, which also illustrates the fact that imidazole is a good nucleophile. However, no activation is necessary with 2-halo-1,3-azoles, which can react with nucleophiles, as shown by the preparations of **3.42** and **3.43**.

Once again this reactivity parallels certain features of carbonyl chemistry. Compare the reaction of aniline with chloroformates, below.

3.42

3.43

3.8 Problems

1. Suggest a synthesis of oxazole **3.33**.

3.33

2. A less general synthesis of oxazoles is the condensation of bromoketones with amides. What is the mechanism for the formation of oxazole **3.44**? How does **3.44** relate to the oxazole which might be prepared from the same bromoketone by conversion to the corresponding aminoketone, N-formylation, and cyclocondensation?

3.44

3. Carboxylic acid **3.46** has been extensively used in the preparation of semi-synthetic penicillins and cephalosporins. Devise a synthesis of **3.46** from ester **3.45**.

3.45   **3.46**

3.9 References

Campbell, M.M. (1979). In *Heterocyclic chemistry (*ed. P.G. Sammes) (Vol. 4 of *Comprehensive organic chemistry*, ed. D. Barton and W.D. Ollis), p. 962 (oxazoles) and p. 967 (thiazoles). Pergamon Press, Oxford.

Furniss, B.S., Hannaford, A.J., Smith, P.W.G., and Tatchell, A.R. (1989). *Vogel's textbook of practical organic chemistry* (5th edn), p.1153 (preparation of aminothiazole **3.32**). Longman, Harlow.

Grimmett, M.R. (1970). *Adv. heterocyclic chem.*, **12**, 103 (imidazoles).

Grimmett, M.R. (1979). In *Heterocyclic chemistry* (ed. P.G. Sammes) (Vol. 4 of *Comprehensive organic chemistry*, ed. D. Barton and W.D. Ollis), p.357 (imidazoles). Pergamon Press, Oxford.

Grimmett, M.R. (1980). *Adv. heterocyclic chem.*, **27**, 241 (imidazoles).

Lakhan, R. and Ternai, B. (1974). *Adv. heterocyclic chem.*, **17**, 99 (oxazoles).

Metzger, J.V. (1979). In *Thiazole and its derivatives (The chemistry of heterocyclic compounds* [ed. A. Weissburger and E.C. Taylor], Vol. 34). Wiley Interscience, New York.

Turchi, I.J. (1986). In *Oxazoles (The chemistry of heterocyclic compounds* [ed. A. Weissburger and E.C. Taylor], Vol. 45). Wiley Interscience, New York.

Turchi, I.J. and Dewar, M.J.S. (1975). *Chem. rev.*, **75**, 389 (oxazoles).

4. Isoxazoles, pyrazoles, and isothiazoles

Isoxazole **4.1**, pyrazole **4.2**, and isothiazole **4.3** are the parent structures of the 1,2-azole family of heterocycles, having a nitrogen atom plus one other heteroatom in a 1,2-relationship in a five-membered ring.

4.1   **4.2**   **4.3**

The aromatic sextet is completed by delocalisation of the lone pair from the second heteroatom, **4.4a–e**. Consequently, as in pyridine, the nitrogen atoms of the 1,2-azoles have a lone pair available for protonation. However the 1,2-azoles are significantly less basic than the 1,3-azoles because of the electron-withdrawing effect of the adjacent heteroatom. Isoxazole and isothiazole are essentially non-basic heterocycles (pK_as <0), and even pyrazole (pK_a=2.5) is a much weaker base than the corresponding 1,3-azole imidazole (pK_a=7).

4.4a   **4.4b**   **4.4c**   **4.4d**   **4.4e**

X = O,NH,S

As with substituted imidazoles, substituted pyrazoles may exist as a mixture of tautomers. 5-Methyl pyrazole **4.5** and 3-methyl pyrazole **4.6** exist as a rapidly equilibrating mixture in solution.

4.5   **4.6**

Although there are a few examples of naturally-occurring 1,2-azoles, many totally synthetic derivatives have found pharmaceutical application.

4.1 Synthesis of isoxazoles and pyrazoles

Retrosynthetic disconnection of generalised 1,2-azole **4.7** gives initially **4.8** which would exist as ketone **4.9**. This in turn is clearly derived from 1,3-diketone **4.10**.

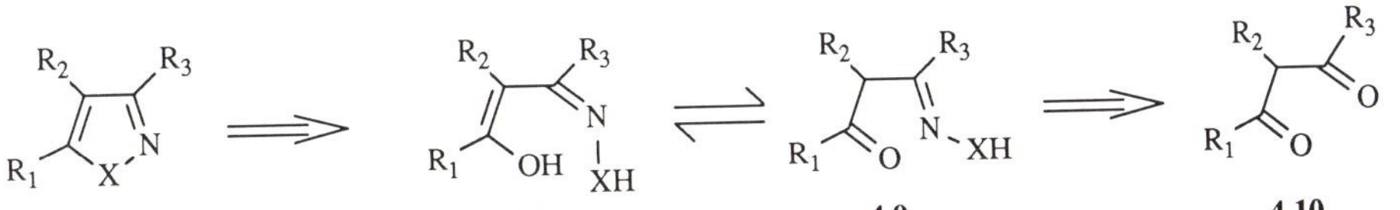

4.11	H_2NOH
4.12	H_2NNH_2
4.13	H_2NSH

X=O,NH,S

In practice hydroxylamine and hydrazine are very reactive nucleophiles, far more so than might be expected from consideration of simple physical parameters. The inceased nucleophilicity of a heteroatom when bonded to a second hereoatom is known as the α effect. For a theoretical rationalisation of the α effect in terms of frontier obitals see Fleming, 1976.

This analysis suggests that condensation of **4.10** with hydroxylamine **4.11**, hydrazine **4.12**, or thiohydroxylamine **4.13** should give the corresponding 1,2-azole. This approach represents an important route to isoxazoles and pyrazoles, but thiohydroxylamine **4.13**, although known, is far too unstable for synthetic purposes. The synthesis of isothiazoles will be mentioned later. The mechanism of the forward process is illustrated by the preparation of isoxazole **4.14** and is simply two consecutive condensations.

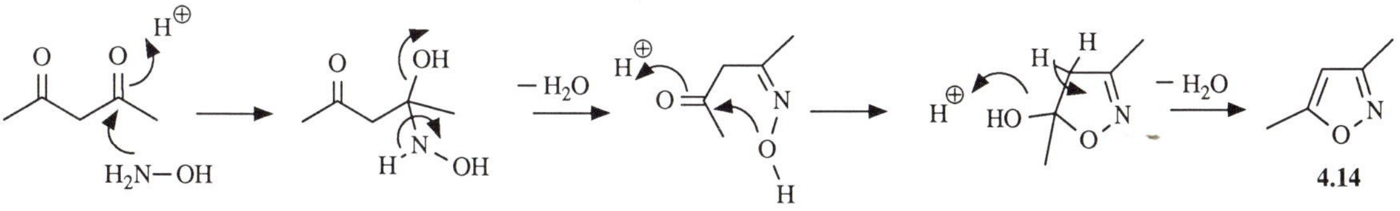

Note that if hydroxylamine or a substituted hydrazine is condensed with an unsymmetrical diketone (**4.10**, where R_1 and R_3 are different) then a regioisomeric mixture of isoxazoles or pyrazoles may result. However a single regioisomer may predominate where there is an inherent bias.

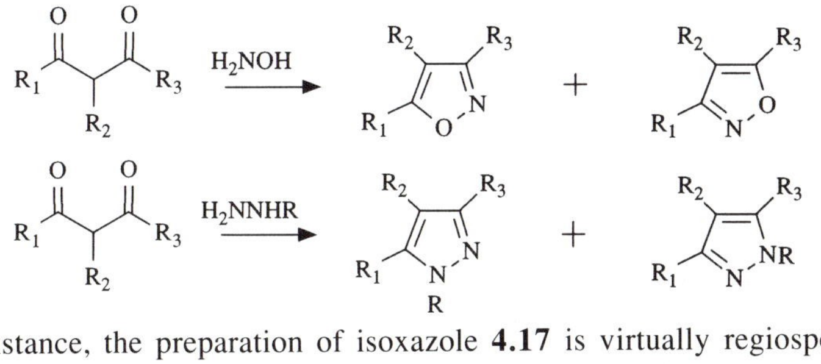

The general reactions of H_2NOH and H_2NNHR with unsymmetrical diketones are shown here.

For instance, the preparation of isoxazole **4.17** is virtually regiospecific because the reaction commences with the more nucleophilic heteroatom (i.e. nitrogen) attacking the more electrophilic ketone (activated by the electron-withdrawing inductive effect of the adjacent ester group). The reader is encouraged to consider the regiochemical bias in the preparation of isoxazole **4.15** and pyrazole **4.16**.

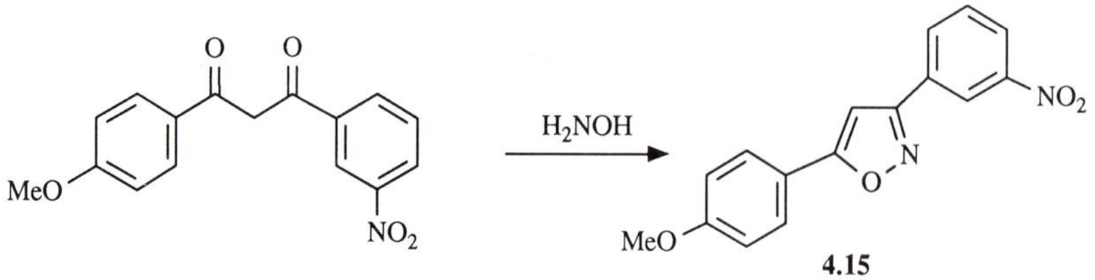

4.15

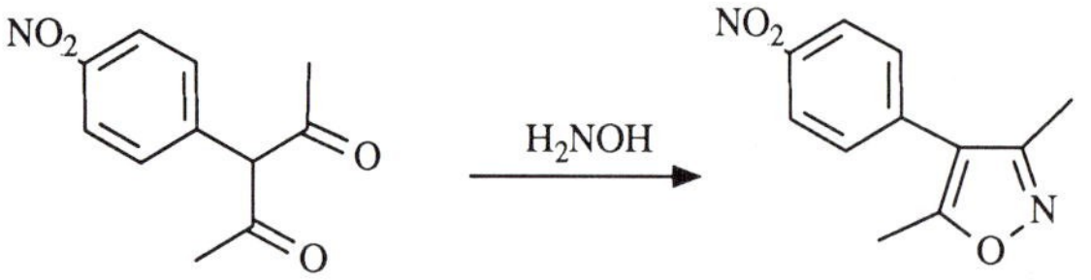

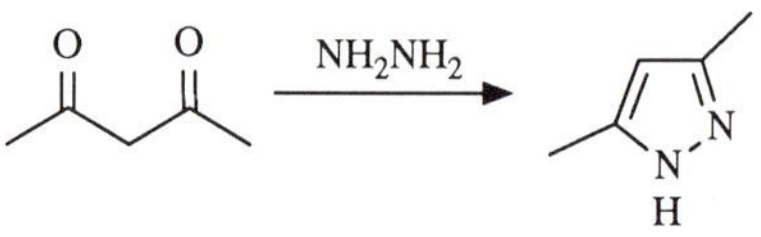

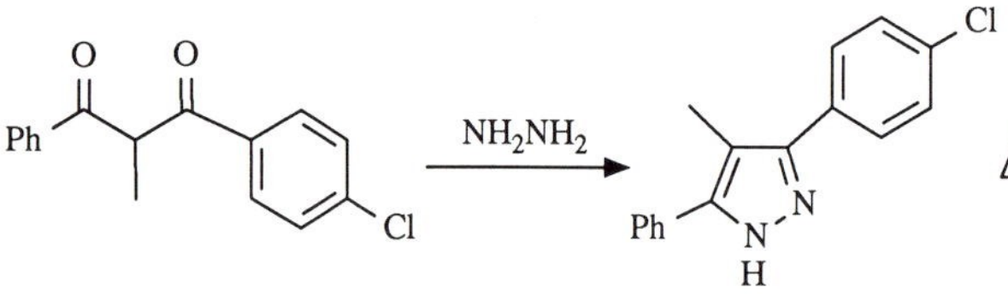

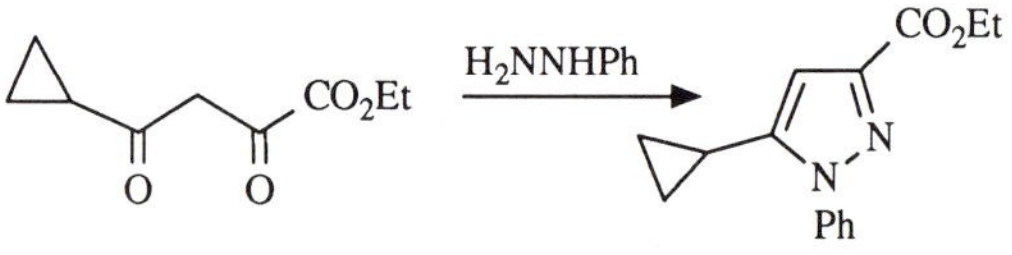

4.16

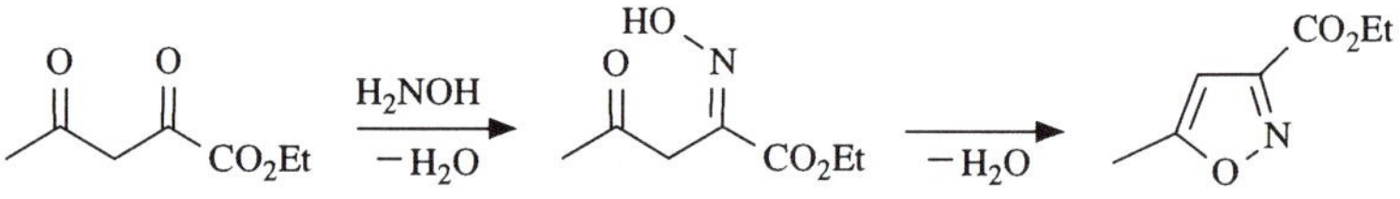

4.17

The other important isoxazole synthesis involves the concerted [3+2] cycloaddition reaction of nitrile oxides **4.18** with either alkynes **4.19** or alkyne equivalents **4.20**.

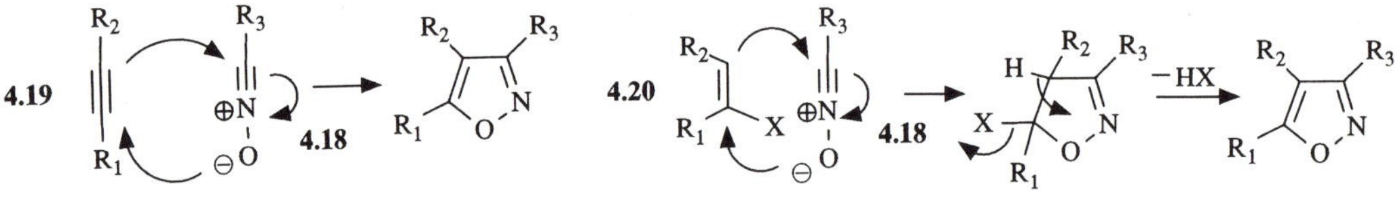

$X=OAc, NMe_2, NO_2$

A wide range of nitrile oxides is known (R_3 = H, aryl, alkyl, ester, halide, etc). The method of choice for the preparation of simple nitrile oxides (R_3 = alkyl, aryl) is oxidation of the corresponding oxime:

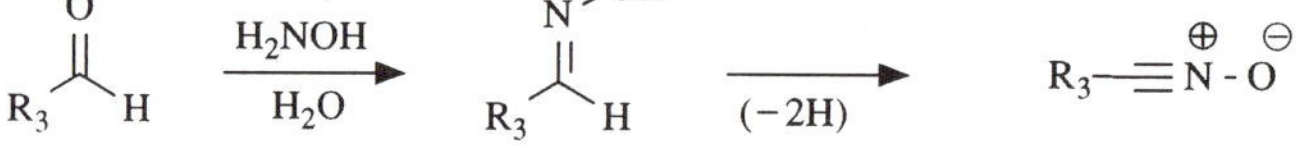

4.18

Several oxidising agents can be used (lead tetraacetate, N-bromosuccinimide, chlorine, etc). A mechanism is illustrated below for alkaline sodium hypochlorite.

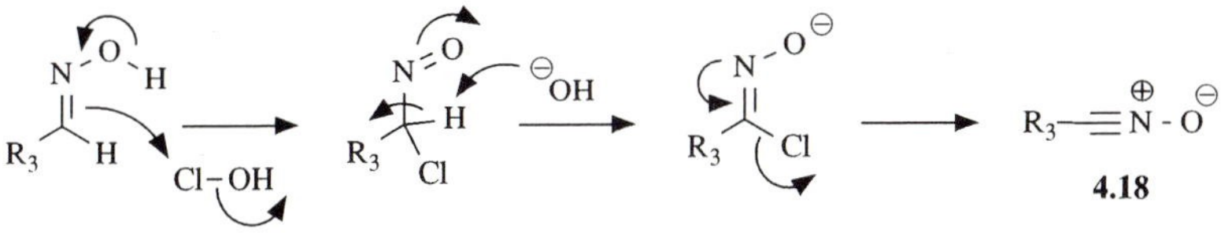

4.18

Let us now consider the synthesis of isoxazole **4.28**, a drug for the treatment of bronchial asthma. The most direct preparation of isoxazolyl ketone **4.24** is the cycloaddition of unstable bromonitrile oxide **4.22** (prepared *in situ* by dehydrobromination of **4.21**) with acetylenic ketone **4.23**. Observe the regioselectivity of this reaction. Both electron-donating and electron-withdrawing groups on the acetylenic components in such cycloadditions tend to occur at the C5 position in the final isoxazole and not at C4. Bromination of ketone **4.24** affords bromoketone **4.25** which is

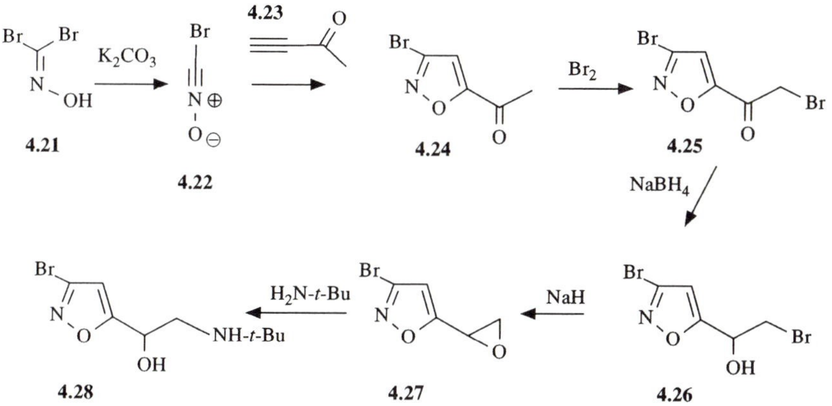

reduced with sodium borohydride to give bromohydrin **4.26**. Treatment with a strong base produces epoxide **4.27** via the intermediate alkoxide, and nucleophilic opening of this epoxide at the least sterically hindered position affords the target drug **4.28**.

4.2 Synthesis of isothiazoles

Isothiazoles are usually prepared by routes involving formation of the nitrogen–sulphur bond in the cyclisation step. This is often set up by oxidation of the sulphur atom, as in the conversion of thioamide **4.29** to isothiazole **4.30**.

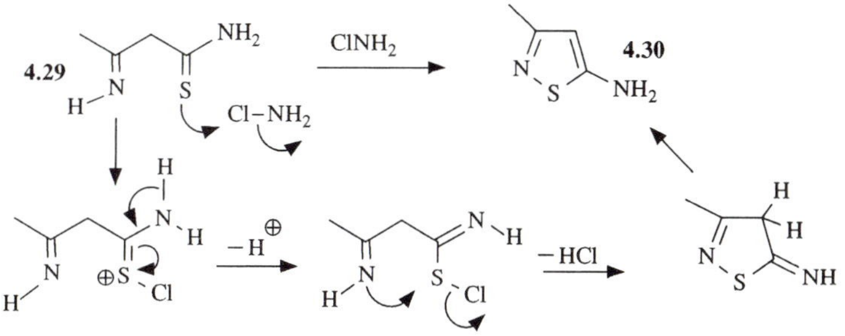

4.3 Electrophilic substitution of isoxazoles, pyrazoles, and isothiazoles

The presence of a pyridine-like nitrogen in the 1,2-azoles makes them markedly less reactive towards electrophilic substitution then furan, pyrrole, and thiophene. (The same effect was noted for the 1,3-azoles in Chapter 3.) Nevertheless, electrophilic substitution is known in 1,2-azoles, occurring principally at the C4 position. This selectivity is reminiscent of pyridine chemistry where the position *meta* to the electronegative nitrogen atom is the 'least deactivated' (see Chapter 5).

Nitration and sulphonation of 1,2-azoles under vigorous conditions are also known, as in the preparation of 4-nitropyrazole **4.31**.

See the related preparation of nitroimidazole **3.36**.

As we have seen with other electron-deficient heterocycles, the introduction of an electron-donating group promotes electrophilic substitution, as in the facile bromination of aminoisothiazole **4.32**.

4.4 Anion chemistry of isoxazoles, pyrazoles, and isothiazoles

Isothiazoles and nitrogen-blocked pyrazoles can be deprotonated at the C5 position with alkyl lithium reagents, and the resultant carbanions quenched with a wide range of electrophiles, as in the preparation of **4.33** and **4.34**.

4.33

4.34

This useful methodology (complementary to the C4 selectivity of normal electrophilic substitution) is not applicable to isoxazole chemistry because the intermediate anions (such as **4.35**) are rather unstable and decompose *via* oxygen–nitrogen cleavage.

4.35

However, alkyl groups at the C5 position of isoxazoles can be deprotonated and reacted with electrophiles.

Note the analogy to the anions derived from crotonoate esters.

Dimethyl isoxazole **4.14** can be selectively deprotonated at the C5 methyl group, nearer the more electronegative oxygen atom. Although simple deprotonation cannot afford an entry into C4 substitution in this system, it is possible to generate a carbanion at the C4 position in a roundabout fashion. Bromination of **4.14** affords the C4-functionalised isoxazole **4.36**. Metal–halogen exchange with *n*-butyllithium at low temperature (-78°C) generates carbanion **4.37** which can be quenched with electrophiles to give isoxazoles such as **4.38**.

Interestingly, 1, 3, 5-trimethyl pyrazole is deprotonated on the N-methyl group, facilitating reaction with electrophiles at this position.

4.14   **4.36**   **4.37**   **4.38**

4.5 Problems

1. What is the mechanism for the formation of isothiazolone **4.39**?

4.39

2. What general strategy might be employed to convert pyrazole to alcohol **4.40**, a potent inhibitor of steroid biosynthesis.

3. What is the product resulting from oxidation of **4.41**?

4.41

4. A synthesis of 2-cyanocyclohexanone **4.45** from cyclohexanone is shown below. Formylation of cyclohexanone produces a mixture of keto/enol tautomers **4.42** and **4.43**, the equilibrium lying to the side of the enol **4.42**. Treatment with hydroxylamine affords isoxazole **4.44**, and base-induced fragmentation of the isoxazole ring affords **4.45**. Explain the regioselectivity of the isoxazole formation, and the mechanism of the fragmentation process.

4.6 References

Campbell, M.M. (1979). In *Heterocyclic chemistry* (ed. P.G. Sammes) (Vol. 4 of *Comprehensive organic chemistry*, ed. D. Barton and W.D. Ollis), p.993 (isoxazoles) and p.1009 (isothiazoles). Pergamon Press, Oxford.

Fleming, I. (1976). *Frontier orbitals and organic chemical reactions*, p.77. Wiley, Chichester.

Furniss, B.S., Hannaford, A.J., Smith, P.W.G., and Tatchell, A.R. (1989). In *Vogel's textbook of practical organic chemistry* (5th edn), p.1149 (preparation of 3,5-dimethylpyrazole). Longman, Harlow.

Grimmett, M.R. (1979). In *Heterocyclic chemistry* (ed. P.G. Sammes) (Vol. 4 of *Comprehensive organic chemistry*, ed. D. Barton and W.D. Ollis), p.357 (pyrazoles). Pergamon Press, Oxford.

Kochetkov, N.K. and Sokolov, S.D. (1963). *Adv. heterocyclic chem.*, **2**, 365 (isoxazoles).

Kost, A.N. and Grandberg, I.I. (1966). *Adv. heterocyclic chem.*, **6**, 347 (pyrazoles).

Slack, R. and Wooldrige, K.R.H. (1965). *Adv. heterocyclic chem.*, **4**, 107 (isothiazoles).

Wakefield, B.J. and Wright, D.J. (1979). *Adv. heterocyclic chem.*, **25**, 147 (isoxazoles).

5. Pyridines

5.1 Introduction

Pyridine **5.1** is a polar liquid (b.p. 115°C) which is miscible with both organic solvents and water. It can formally be derived from benzene by replacement of a CH group by a nitrogen atom. Pyridine is a highly aromatic heterocycle, but the effect of the heteroatom makes its chemistry quite distinct from that of benzene. The aromatic sextet of six π electrons is complete without invoking participation of the lone pair on the nitrogen. This is in direct contrast with the situation in pyrrole (Chapter 2) where the aromatic sextet includes the lone pair on the nitrogen. Hence the lone pair of pyridine is available for bonding without disturbing the aromaticity of the ring. Pyridine is moderately basic (pK_a=5.2) and can be quaternised with alkylating agents to form pyridinium salts **5.2**. Pyridine also forms complexes with Lewis acids such as sulphur trioxide. This complex **5.3** is a mild source of sulphur trioxide for sulphonation reactions (see Chapter 2).

The effect of the heteroatom is to make the pyridine ring very unreactive to normal electrophilic aromatic substitution. Conversely, pyridines are susceptible to nucleophilic attack. These topics are discussed later.

5.2 Synthesis of pyridines

Our retrosynthetic analysis of generalised pyridine **5.4** commences with an adjustment of the oxidation level to produce dihydropyridine **5.5**. This molecule can now be disconnected very readily. Cleavage of the carbon–heteroatom bonds in the usual way leaves dienol **5.6** which exists as diketone **5.7**. The 1,5-dicarbonyl relationship can be derived from a Michael reaction of ketone **5.8** and enone **5.9**, which in turn can arise from condensation of aldehyde **5.10** and ketone **5.11**.

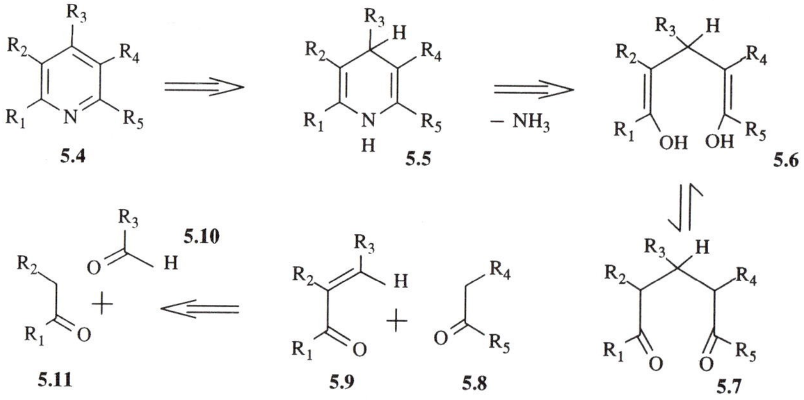

These processes are facilitated when R_2 and R_4 are electron-withdrawing groups such as esters. Furthermore, when ketones **5.11** and **5.8** are the same, we have the basis for the classical Hantzsch pyridine synthesis.

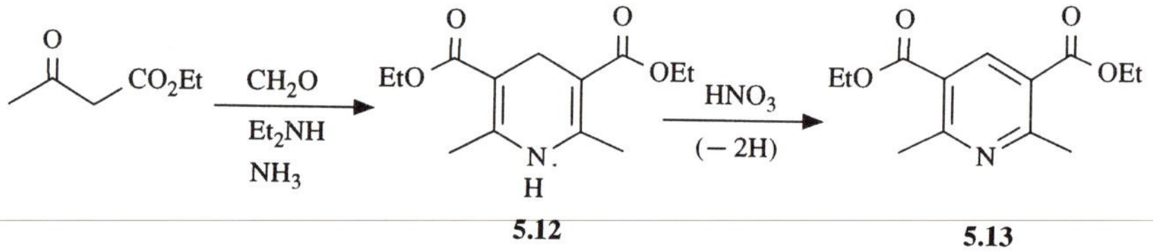

For instance, condensation of ethyl acetoacetate, formaldehyde, and ammonia gives dihydropyridine **5.12** which is readily oxidised with nitric acid to give pyridine **5.13**. Although the precise details of this multicomponent condensation are not known, a reasonable pathway is shown below.

Note that in this example R_2 and R_4 are ethyl esters, so the adjacent carbon is actually an active methylene group. The higher acidity and hence nucleophilicity of these centres facilitates the reaction sequence.

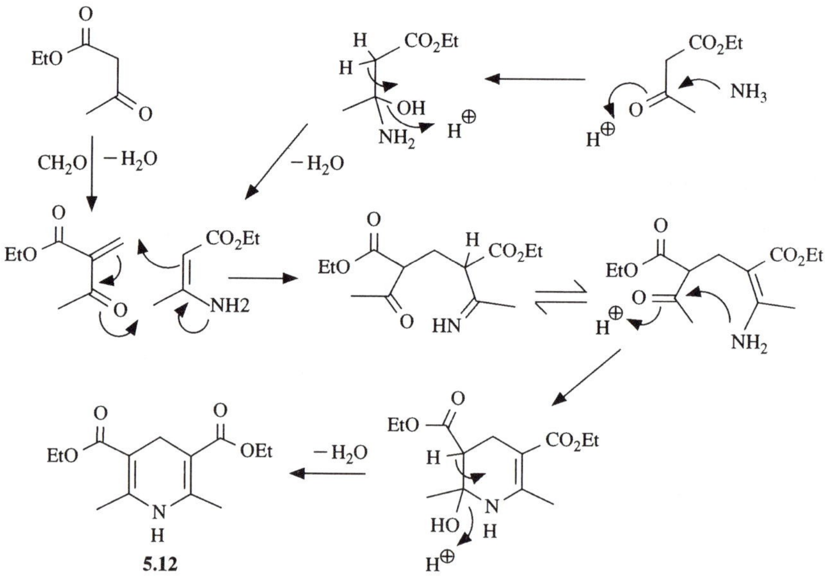

Some examples of dihydropyridines prepared in this way are shown below. (The student is encouraged to work out the aldehydes used in each case.)

As well as being intermediates for the synthesis of pyridines, these dihydropyridines are themselves an important class of heterocycles. For instance, dihydropyridine **5.14** is a drug for lowering blood pressure. In the synthesis of **5.14** note that carrying out the Hantzsch synthesis stepwise allows for the preparation of an unsymmetrical dihydropyridine, having both a methyl and an ethyl ester.

A consequence of the asymmetry of **5.14** is that C4 is a chiral centre. Hence the product is formed as a racemic mixture.

5.3 Electrophilic substitution of pyridines

Pyridine is virtually inert to aromatic electrophilic substitution. Consider nitration of pyridine by nitric acid. First, as pyridine is a moderate base, it will be almost completely protonated by the acid, making it much less susceptible to electrophilic attack. Second, addition of the electrophile to the small amount of unprotonated pyridine present in solution is not a facile process.

Attack of the electrophile at the C2 or C4 position results in an intermediate cation with partial positive charge on the electronegative nitrogen atom. This is clearly not energetically favourable when compared to C3 substitution, where no partial positive charge resides on nitrogen. In fact the product of C3 substitution, nitropyridine **5.15**, can be isolated from the exhaustive nitration of pyridine, but only in poor yield.

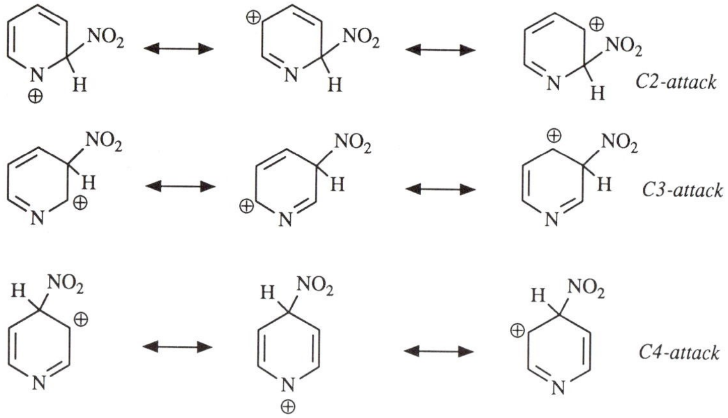

C2-attack

C3-attack

C4-attack

C-alkylation of a sterically-hindered phenolate anion.

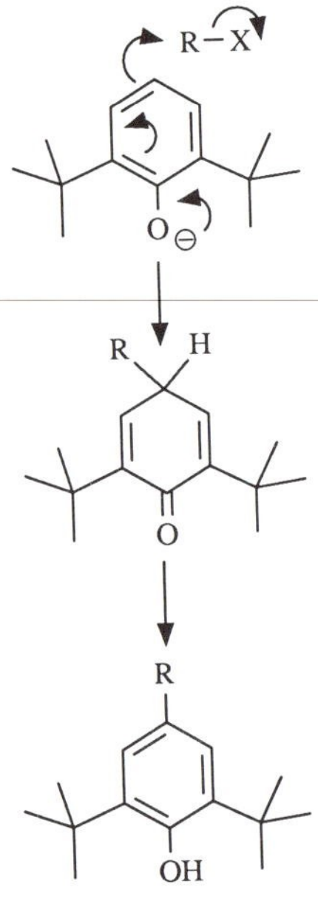

Although better results have been achieved with the sulphonation of pyridine to give the sulphonic acid **5.16**, electrophilic substitutions on an inactivated pyridine ring are in general not preparatively useful.

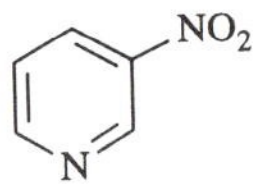

5.15

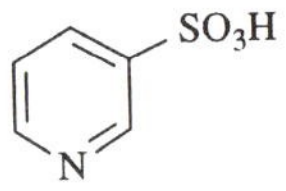

5.16

Pyridine can be activated to electrophilic substitution by conversion to pyridine N-oxide **5.17**. At first sight it is curious to consider oxidation (i.e. electron loss) as a means of activating a system to electrophilic substitution, but **5.17** can act rather like sterically-hindered a phenolate anion towards electrophiles, producing intermediate **5.18** which then loses a proton to give substituted N-oxide **5.19**. For this methodology to be useful it is of course necessary to remove the activating oxygen atom. This can be done with phosphorous trichloride, which becomes oxidised to phosphorous oxychloride.

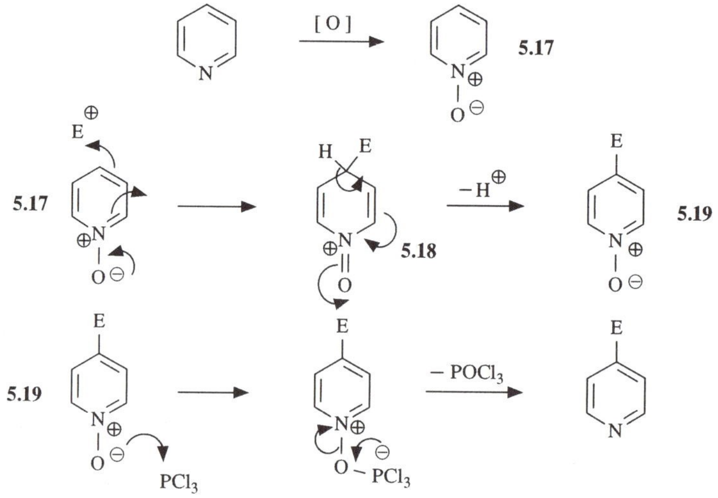

For instance, 4-nitropyridine **5.20** can be prepared from pyridine in three steps by this methodology.

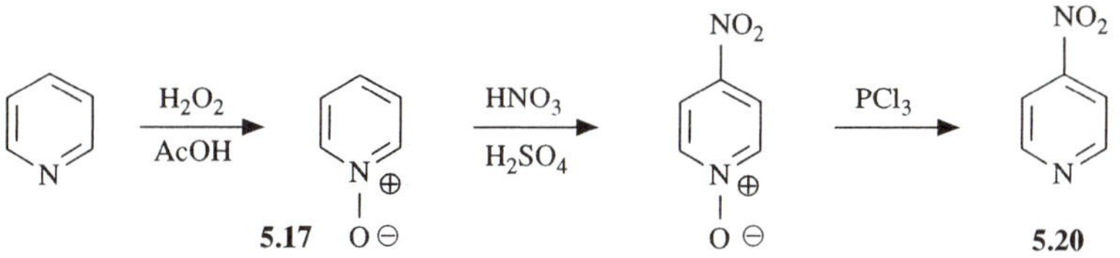

Pyridine N-oxides can also be converted into synthetically useful 2-chloropyridines **5.21** (see later).

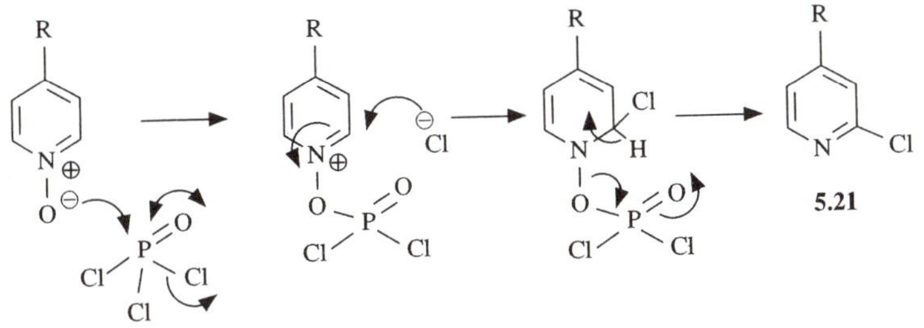

Another approach to electrophilic substitution involves the chemistry of 2-pyridone **5.22** and 4-pyridone **5.23**. These are the tautomeric forms of 2- and 4-hydroxypyridine respectively. They exist exclusively in the pyridone form, the hydrogen atom being attached to the nitrogen atom, not the oxygen. Their electronic structures are not adequately described by a single valence representation, the lone pair from the nitrogen atom being delocalised to a considerable extent onto the oxygen atom, as in mesomeric representations **5.22a** and **5.23a**.

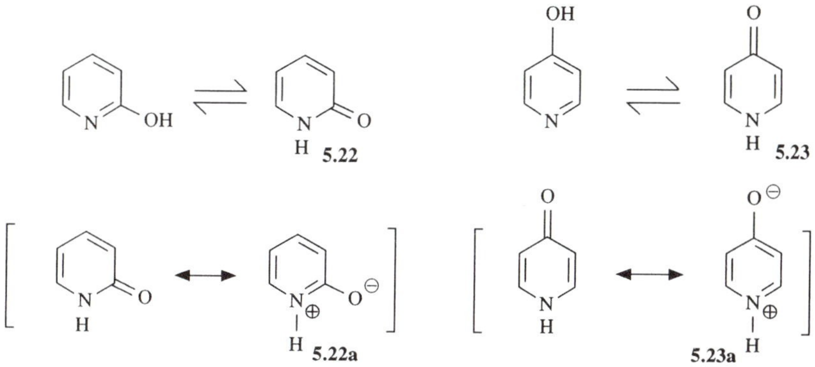

Both pyridones can react with electrophiles at positions *ortho* and *para* to the activating oxygen atom. For instance, 4-pyridone reacts with electrophiles at the C3 position (the mechanism can be formulated from either mesomeric representation) to give intermediate **5.24**. As with pyridine N-oxides, reaction with phosphorous oxychloride gives useful chloropyridines **5.25**. We shall see the utility of 2- and 4-chloropyridines in the next section.

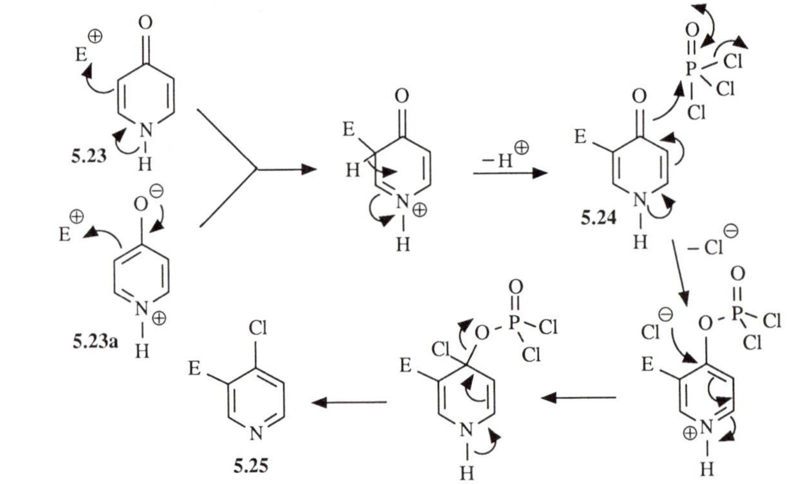

5.4 Nucleophilic substitution of pyridines

Pyridine can be attacked by nucleophiles at the C2/C6 and C4 positions in a manner analogous to the addition of nucleophiles to a carbonyl group in a 1,2 or 1,4 fashion. Attack at the C3/C5 positions is not favoured because the negative charge on the intermediate cannot be delocalised onto the electronegative nitrogen atom.

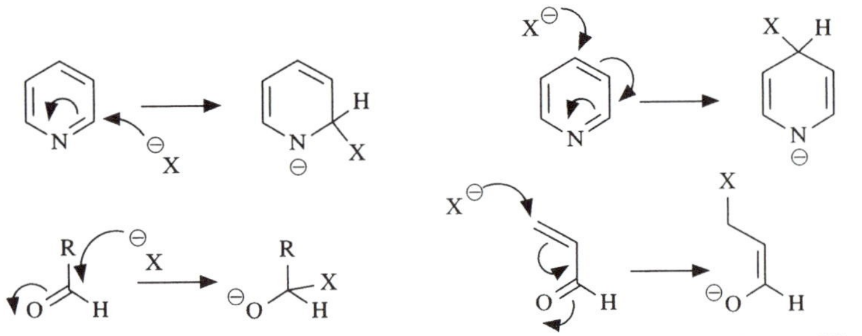

X = Nucleophile

The actual mechanism is rather complicated. Hydrogen gas is evolved, but in reality free sodium hydride is never generated. See McGill and Rappa (1988).

Under conditions of high temperatures the intermediate anion can re-aromatise by loss of a hydride ion, even though it is a very poor leaving group. This is illustrated by the Chichibabin reaction of pyridine and sodamide to produce 2-aminopyridine **5.26**. The immediate product of the reaction is **5.27**, the sodium salt of **5.26**, as the eliminated hydride ion is very basic. Protonation of this sodium salt during the aqueous workup then regenerates **5.26**. A simplistic rationale is shown below.

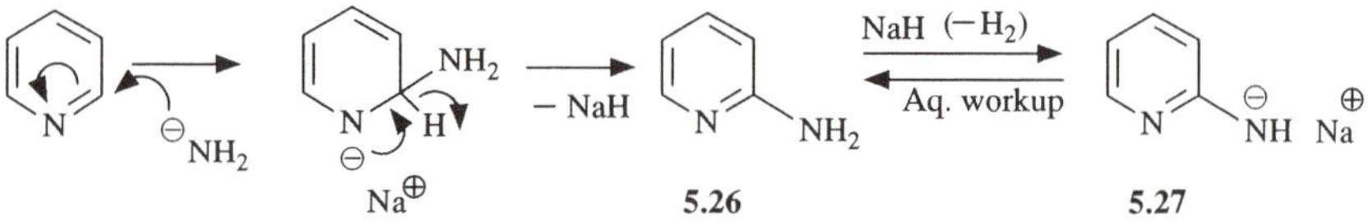

These nucleophilic substitution reactions are much more facile when better leaving groups (e.g. halide ions instead of hydride ions) are employed.

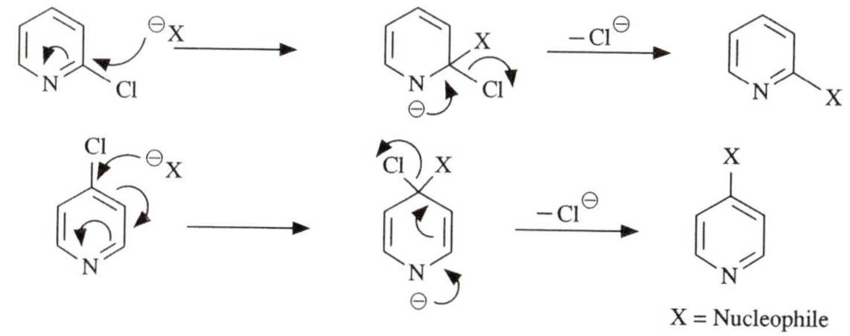

X = Nucleophile

Nucleophilic substitutions are widely used in pyridine chemistry. Some examples are shown below.

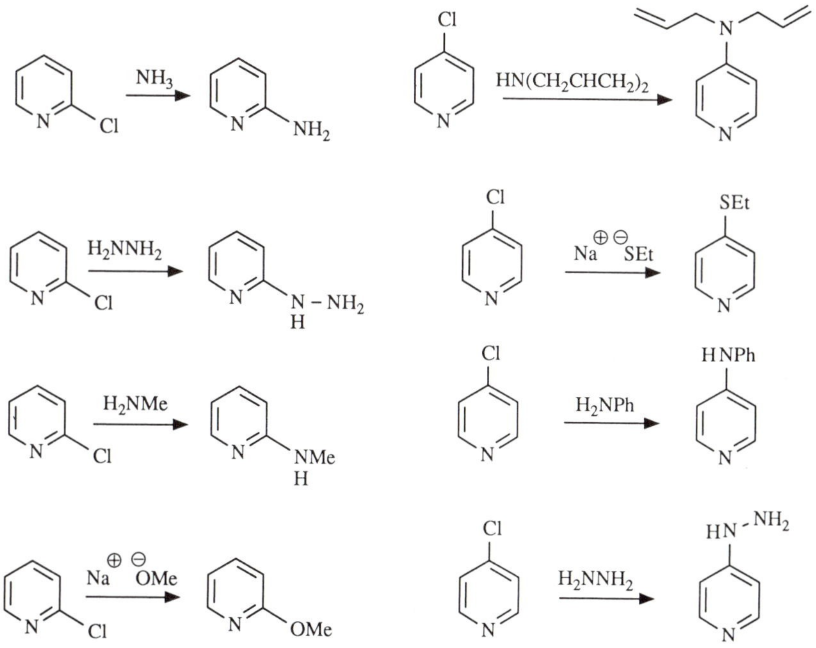

Finally, before leaving this section, we shall consider the synthesis of pyridotriazine **5.32**, a potential anti-fungal drug. This synthesis illustrates features of both electrophilic and nucleophilic pyridine chemistry.

Nitration of 4-pyridone **5.23** gives **5.28**, and reaction with phosphorous oxychloride affords chloropyridine **5.29**. This pyridone-chloropyridine conversion activates the system to nucleophilic attack by hydrazine, affording **5.30**. The nitro group also facilitates nucleophilic attack by delocalisation of negative charge in the intermediate.

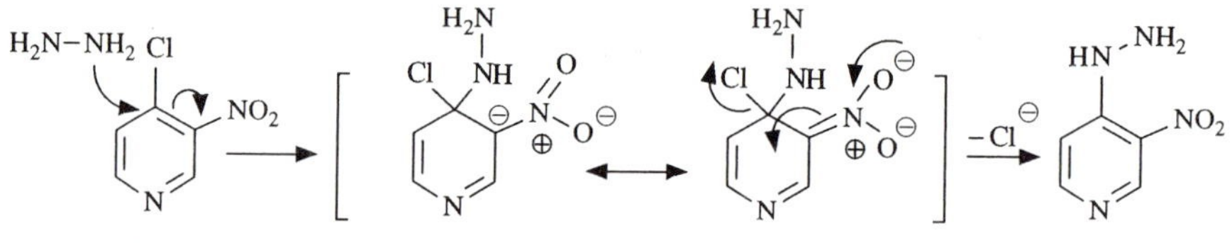

5.29   **5.30**

N-Acylation, reduction of nitro to amino, and condensation produce dihydrotriazine **5.31**. This system is readily dehydrogenated with manganese dioxide to afford the fully aromatic heterocycle **5.32**. Note how relatively simple chemistry can be used to form a quite complex heterocycle.

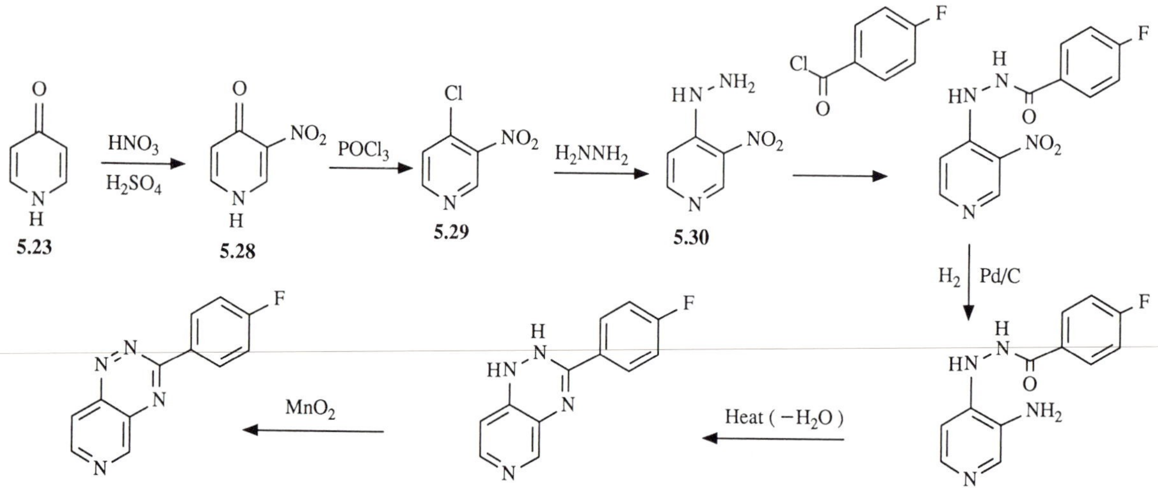

5.5 Anion chemistry of pyridines

We earlier drew a parallel between nucleophilic attack on the C2/C6 and C4 positions of pyridine and 1,2 and 1,4 addition of nucleophiles to a carbonyl group. This analogy can be extended to deprotonation of alkyl substituents at the C2/C6 and C4 positions.

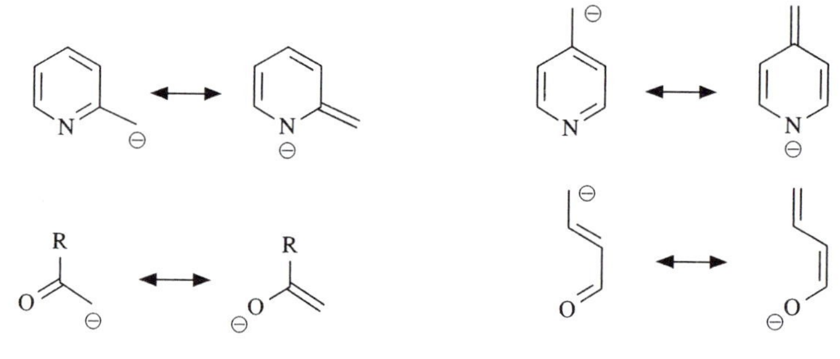

Just as a carbonyl group stabilises an adjacent negative charge as an enolate anion, so the anion derived from 2-methyl pyridine is stabilised by delocalisation of the negative charge onto the electronegative nitrogen atom.

A similar argument holds for 4-methyl pyridine. These stabilised anions can then react with the usual range of electrophiles.

The negative charge resulting from deprotonation of the ethyl methylene group of **5.33** cannot be delocalised onto the nitrogen atom.

Dialkyl pyridine **5.33** is selectively deprotonated at the C4 alkyl group, illustrating the greater acidity of this position over the C3 position. With regard to ring deprotonation, however, there are relatively few examples known for simple pyridines, in contrast to the extensive chemistry developed for the five-membered ring heterocycles. This is because the resultant organometallic species are good nucleophiles, and because pyridines are also moderate electrophiles, polymerisation problems are often encountered. More success has been achieved with substituted pyridines having an *ortho* activating substituent (e.g. -CONHR, -NHCOR, -OMe, -CH$_2$NR$_2$ etc). These substituents increase the rate of kinetic deprotonation and stabilise the intermediate organolithium species by coordination.

For instance, 4-aminopyridine **5.34** can be converted to amide **5.35** which, on treatment with two equivalents of butyl lithium, gives organometallic species **5.36**. Formylation of the more reactive anion (the carbanion) then re-protonation of the amide anion gives **5.37**. Acidic hydrolysis removes the activating group to release pyridine aldehyde **5.38**.

The metalation proceeds by initial deprotonation of the amide followed by *ortho*-directed deprotonation at the C3 position to produce the pseudo six-membered ring organolithium species **5.36**.

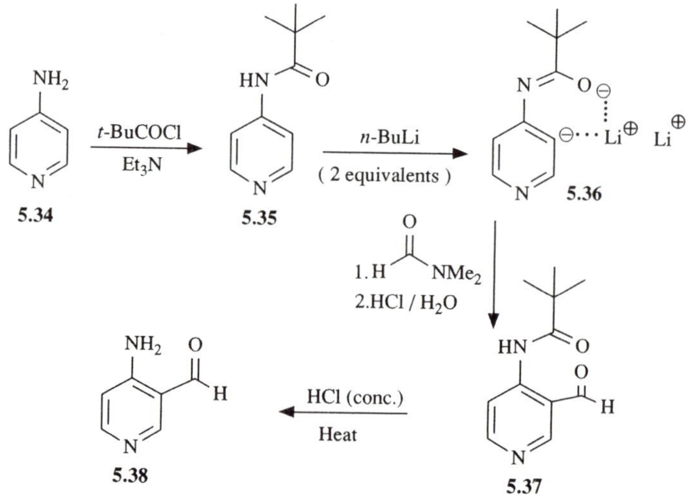

5.6 Problems

1. What is the mechanism of this reaction?

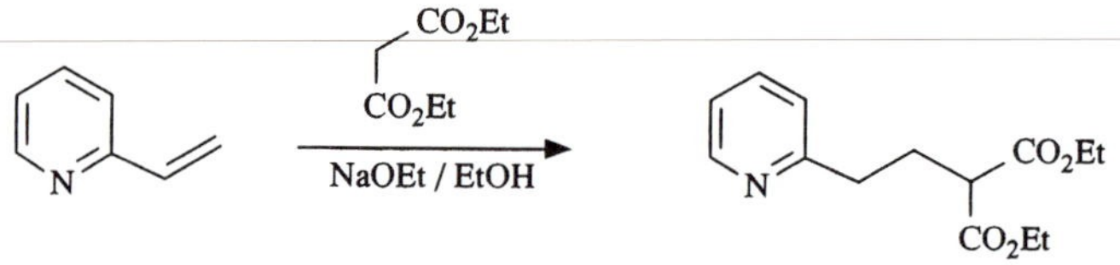

Hint. Start by acetylating the pyridine to give a quaternary cationic species. How can deprotonation afford a nucleophilic enamine-like system?

2. The condensation of active methyl groups with aldehydes can be catalysed with acetic anhydride as well as base. Suggest a possible mechanism.

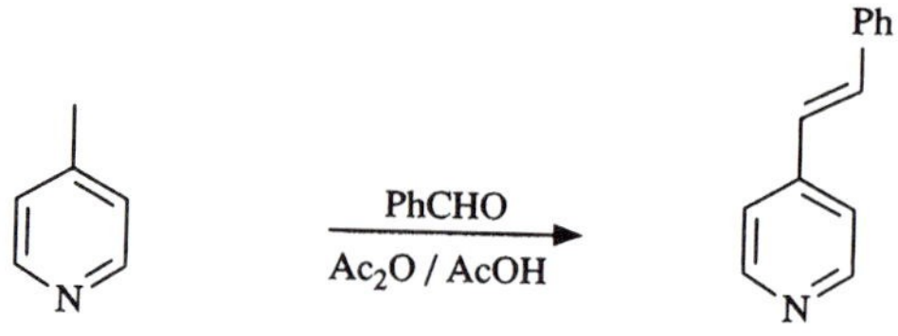

3. Rationalise the formation of lactone **5.40** from pyridyl amide **5.39**.

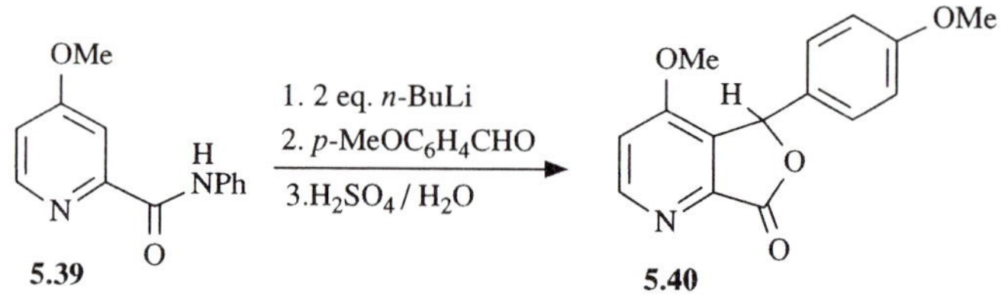

4. Some pyridine N-oxides are not just synthetic intermediates, but are of interest in their own right. For instance, pyridine N-oxide **5.41** is a new drug

claimed to be useful for the treatment of senile dementia. What are the mechanisms of the pyridone-forming step and the final displacement?

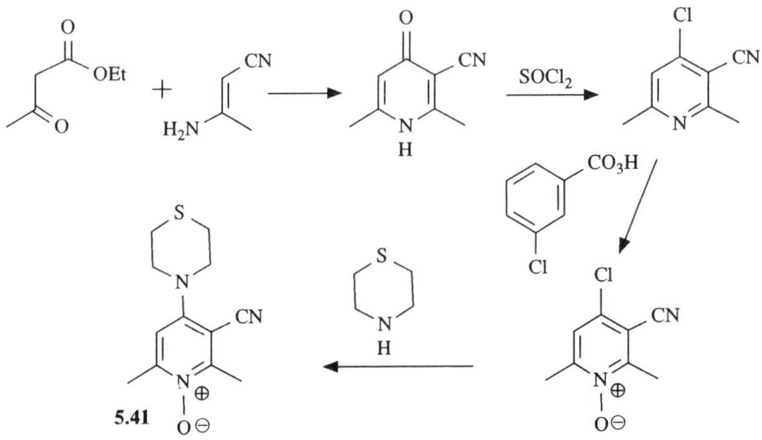

5.7 References

Abramovitch, R.A. (1974). In *Pyridine and its derivatives (The chemistry of heterocyclic compounds* [ed. A. Weissburger and E.C. Taylor], Vol. 14, Supplement Parts 1 – 4). Wiley Interscience, New York.

Eisner, V. and Kuthum, J. (1972). *Chem. rev.*, **72**, 1 (dihydropyridines).

Furniss, B.S., Hannaford, A.J., Smith, P.W.G., and Tatchell, A.R. (1989). *Vogel's textbook of practical organic chemistry* (5th edn), p.1168 (preparation of pyridine **5.13**). Longman, Harlow.

Klinsberg, E. (1974). In *Pyridine and its derivatives (The chemistry of heterocyclic compounds* [ed. A. Weissburger and E.C. Taylor], Vol. 14, Parts 1 – 4). Wiley Interscience, New York.

McGill, C.K. and Rappa, A. (1988). *Adv. heterocyclic chem.*, **44**, 3.

Smith, D.M. (1979). In *Heterocyclic chemistry* (ed. P.G. Sammes) (Vol. 4of *Comprehensive organic chemistry*, ed. D. Barton and W.D. Ollis), p.3. Pergamon Press, Oxford.

6. Quinolines and isoquinolines

6.1 Introduction

Quinoline and isoquinoline can also be viewed as being formally derived from naphthalene

Quinoline **6.1** and isoquinoline **6.2** are two isomeric heterocyclic systems, which can be envisaged as being constructed from the fusion of a benzene ring at the C2/C3 and C3/C4 positions of pyridine respectively. They are both ten π-electron aromatic heterocycles. Like pyridine, they are moderately basic (pK_a quinoline = 4.9, pKa isoquinoline = 5.1). Indeed quinoline is sometimes used as a high boiling-point (237°C) basic solvent.

Note the numbering system for isoquinoline

As with pyridine, the nitrogen atoms of quinoline and isoquinoline each bear a lone pair of electrons not involved in aromatic bonding which can be protonated, alkylated, or complexed to Lewis acids. This chapter should be read in conjunction with the chapter on pyridines as several points discussed at length there are also relevant to the chemistry of quinoline and isoquinoline.

6.2 Synthesis of quinolines and isoquinolines

The classical Skraup synthesis of quinolines is exemplified by the reaction of aniline **6.3** with glycerol **6.4** under acidic/oxidative conditions to produce quinoline **6.1**.

At first sight this reaction appears to be another one of those ancient heterocyclic syntheses that owe more to alchemy than to logic, but in fact the processes involved are relatively straightforward.

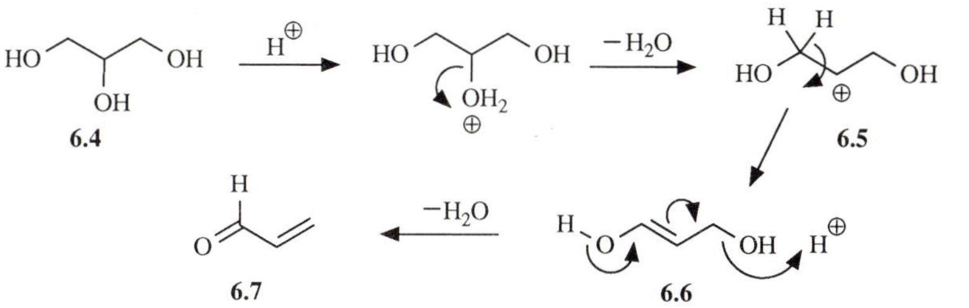

Protonation of glycerol **6.4** catalyses dehydration *via* secondary carbonium ion **6.5** to give enol **6.6**. Acid catalysed elimination of a second water molecule affords acrolein **6.7**. Thus glycerol acts essentially as a protected form of acrolein, slowly releasing this unstable α,β-unsaturated aldehyde into the reaction medium. Better yields are realised with this approach than if acrolein itself is present from the start. The reaction proceeds with a Michael addition of aniline **6.3** to acrolein, producing saturated aldehyde **6.8** which cyclises *via* an aromatic substitution reaction to alcohol **6.9**. Acid-catalysed dehydration to **6.10** then oxidation yields quinoline **6.1**. Nitrobenzene can be used as a mild oxidant, as can iodine and ferric salts.

Acrolein is a highly reactive olefin that is prone to polymerisation.

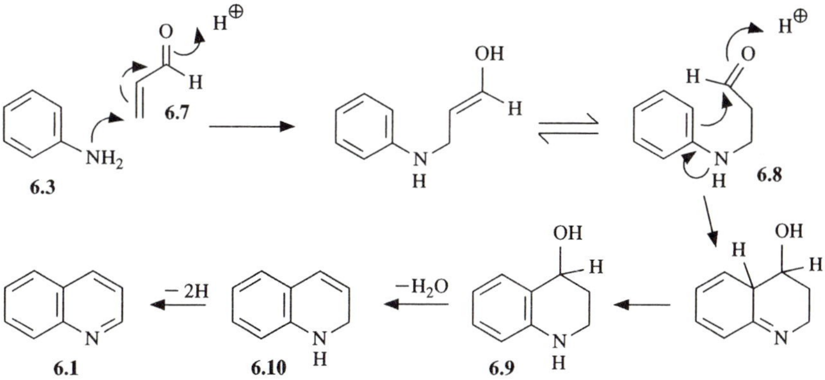

Some examples of the Skraup synthesis are shown below.

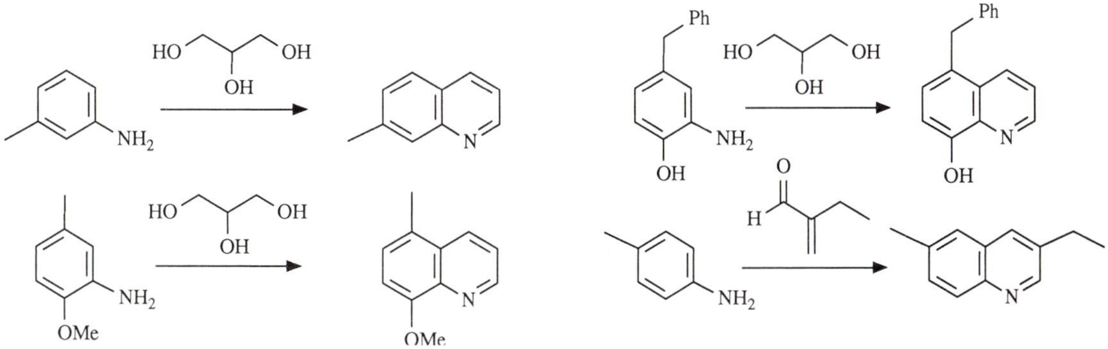

The key intermediates in the synthesis of isoquinolines are β-arylethylamines. For instance, acylation of β-phenylethylamine **6.11** gives amides of general structure **6.12** which can be cyclised with phosphorous oxychloride to produce dihydroisoquinoline **6.13**. Better yields are obtained with electron-donating groups on the aromatic ring facilitating this aromatic substitution cyclisation.

X = Electron-donating substituent

6.11   **6.12**   **6.13**   R   **6.14**   R

This dehydrogenation is the reverse of a normal hydrogenation reaction. The dehydrogenation can be carried out under milder conditions when a hydrogen acceptor (such as cyclohexene) is present.

As in the Skraup quinoline synthesis, loss of two hydrogen atoms is necessary to reach the fully aromatic system. However, this is usually accomplished in a separate step, utilising palladium catalysis to give generalised isoquinoline **6.14**. This is known as the Bischler–Napieralski synthesis. The mechanism probably involves conversion of amide **6.12** to protonated imidoyl chloride **6.15** followed by electrophilic aromatic substitution to give **6.13**. (For a similar activation of an amide to an electrophilic species see the Vilsmeier reaction, Chapter 2.)

6.15   **6.13**

X = Electron-donating substituent

The Pictet–Spengler synthesis is usually used when the tetrahydroisoquinoline oxidation level is required.

Closely related to the Bischler–Napieralski synthesis is the Pictet–Spengler synthesis, which utilises aldehydes rather than acylating species. Condensation of β-arylethylamines with aldehydes produces imines such as **6.16** which can be cyclised with acid to give tetrahydroisoquinoline **6.17**. As with the Bischler–Napieralski synthesis, electron-donating groups (typically methoxy groups) facilitate the cyclisation step. The lower oxidation state of **6.17** as compared to **6.13** is a direct consequence of using a carbonyl group at the aldehyde rather than carboxylic acid oxidation level. Four hydrogen atoms have to be removed from tetrahydroisoquinolines by oxidation to produce the fully aromatic isoquinoline.

6.17   **6.16**

6.3 Electrophilic substitution of quinoline and isoquinoline

Quinoline and isoquinoline undergo electrophilic substitution reactions more easily than pyridine, though not surprisingly the incoming electrophile attacks the benzenoid ring. As with pyridine, the nitrogen atoms of quinoline and isoquinoline are protonated under the typically acidic conditions of nitration or sulphonation, making the heterocyclic ring resistant to attack. The C5 and C8 positions are most susceptible to electrophilic attack.

Attack of an electrophile at C5 of protonated quinoline gives cation **6.18a,b** which is stabilised by resonance as shown without disturbing the aromaticity of the adjacent pyridinium ring. However, attack of an electrophile at C6 produces cation **6.19** which does not possess the same resonance stabilisation of cation **6.18a,b**. (The student should perform the same exercise for the C7 and C8 positions and confirm that the same arguments can be applied.)

For instance, nitration of quinoline gives an equal mixture of regioisomers **6.20** and **6.21**. However, nitration of isoquinoline is reasonably selective (10:1) for the C5 position over the C8, affording mainly **6.22**.

6.4 Nucleophilic substitution of quinoline and isoquinoline

Quinoline and isoquinoline undergo nucleophilic substitution reactions, like pyridine.

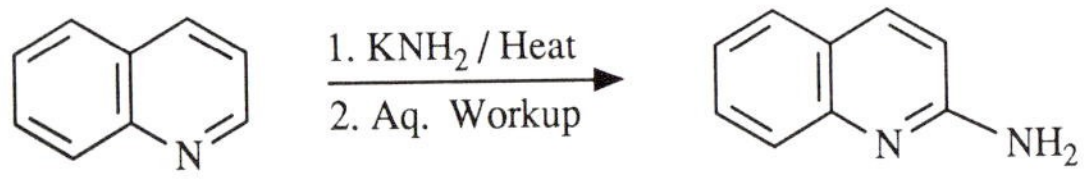

For instance, both quinoline and isoquinoline undergo the Chichibabin reaction (with formal hydride elimination, see Chapter 5) to give 2-aminoquinoline **6.23** and l-aminoisoquinoline **6.24** respectively. Halogen substituents *ortho* to the nitrogen atoms are easily displaced, as in the preparations of **6.25** and **6.26**.

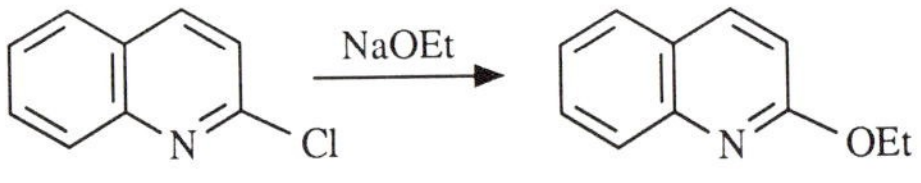

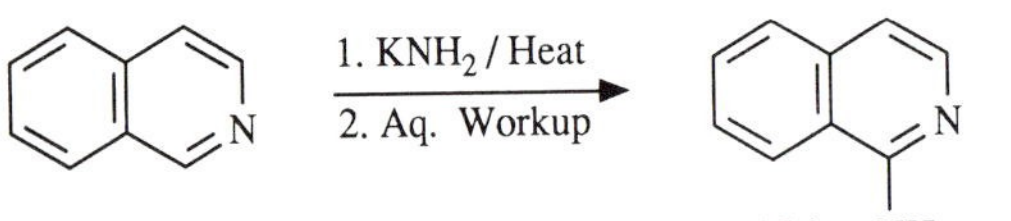

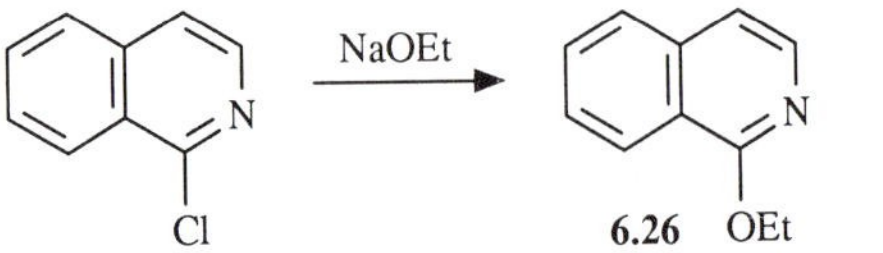

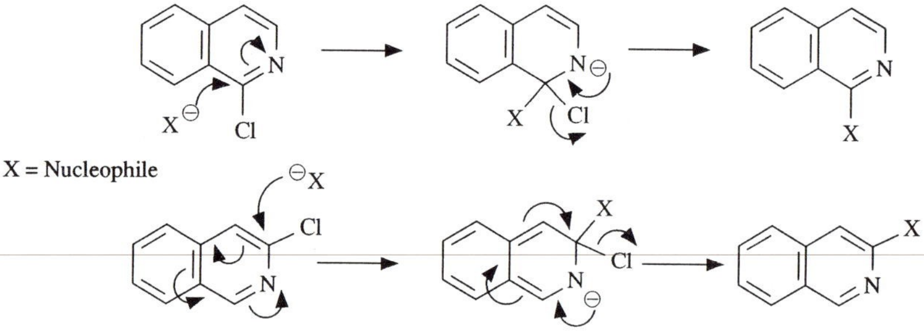

X = Nucleophile

Note that nucleophilic displacement in isoquinolines occurs more easily at the C1 position than at the C3 position (even though they are both *ortho* to nitrogen) because displacement at C3 involves temporary disruption of the benzenoid ring.

6.5 Anion chemistry of quinoline and isoquinoline

Alkyl groups at the C2 and C4 positions of quinoline can be deprotonated by strong bases. This is because (as with pyridine) the negative charge on the resultant carbanions can be delocalised onto the electronegative nitrogen atom, as in carbanion **6.27a,b**.

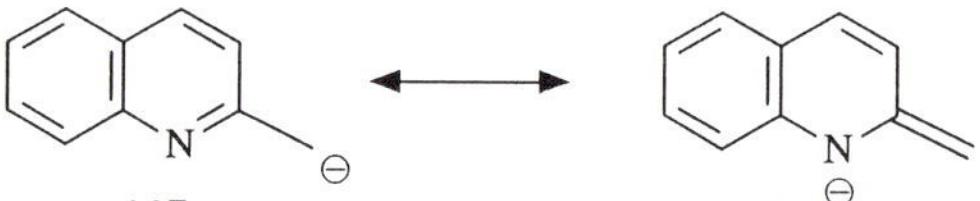

6.27a　　　　　　　　　**6.27b**

Such carbanions can be alkylated, acylated, or condensed with aldehydes:

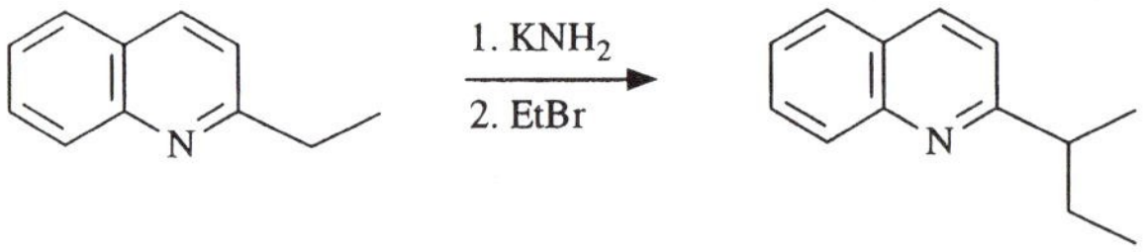

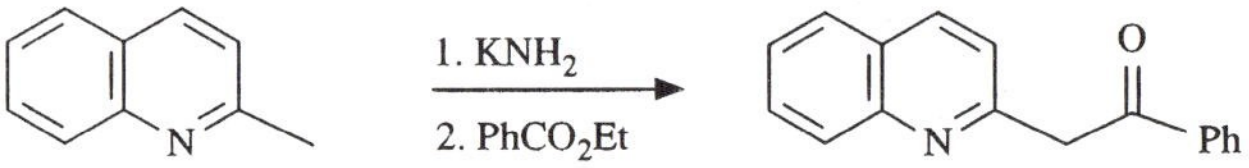

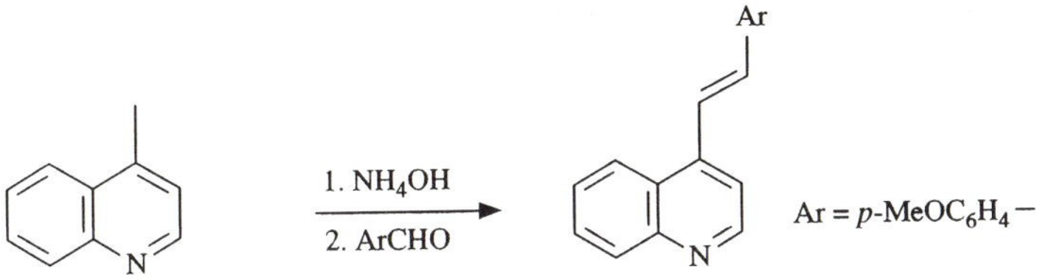

This type of chemistry is also observed with 1-methyl isoquinoline **6.28**. However 3-methyl isoquinoline is much less activated because delocalisation of charge in **6.29a,b** involves disruption of aromaticity of the benzenoid ring. This phenomenon is closely related to the reluctance of 3-halo isoquinolines to undergo nucleophilic substitution.

6.28

6.29a **6.29b**

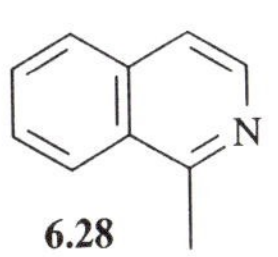

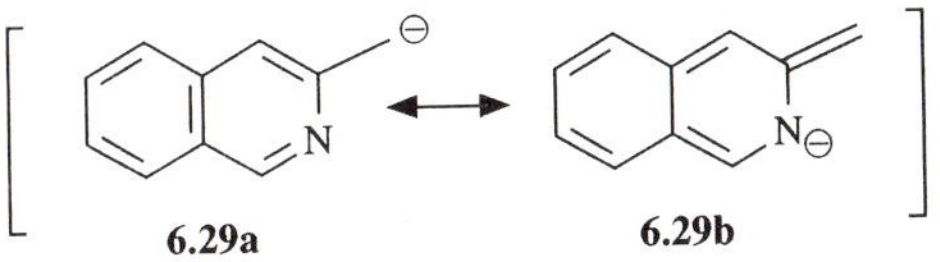

As with pyridine, activated alkyl groups can be condensed with aldehydes under acidic as well as basic conditions, as in the preparation of **6.30** and **6.31.**

6.30

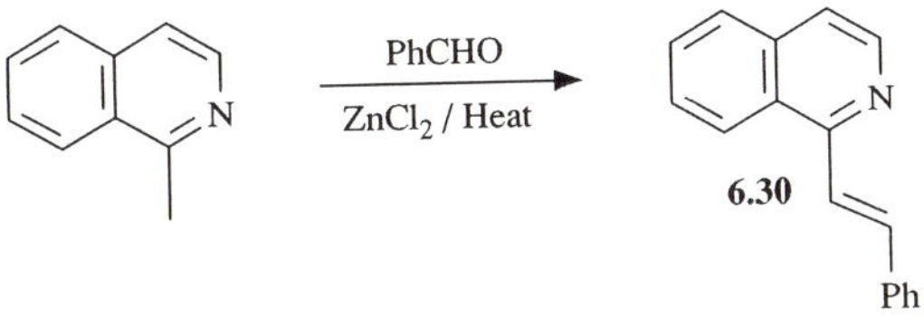

The reader is referred to the previous chapter (Problem 2) for a mechanistic explanation of such condensations.

6.31

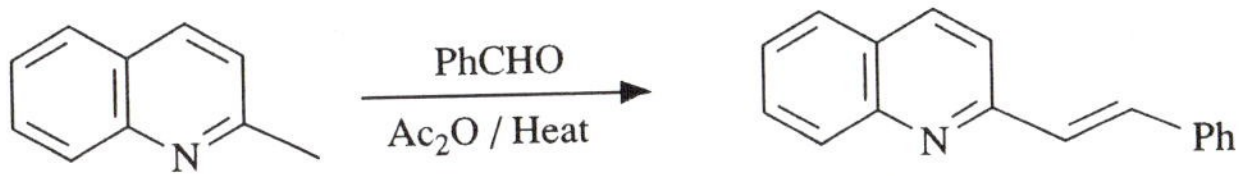

6.6 Problems

1. The synthesis of the important quinolone antibiotic **6.33** is shown. The key stages are the Gould–Jacobson quinolone synthesis to give **6.32**, and the displacement reaction to afford **6.33**. What are the mechanisms of these reactions?

6.32 **6.33**

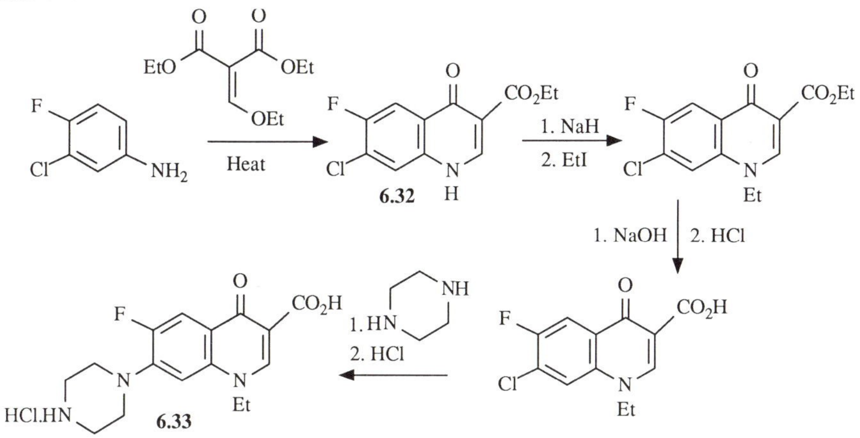

2. A synthesis of the naturally-occurring isoquinoline alkaloid **6.34** is shown below. What reagents might be used to accomplish each transformation?

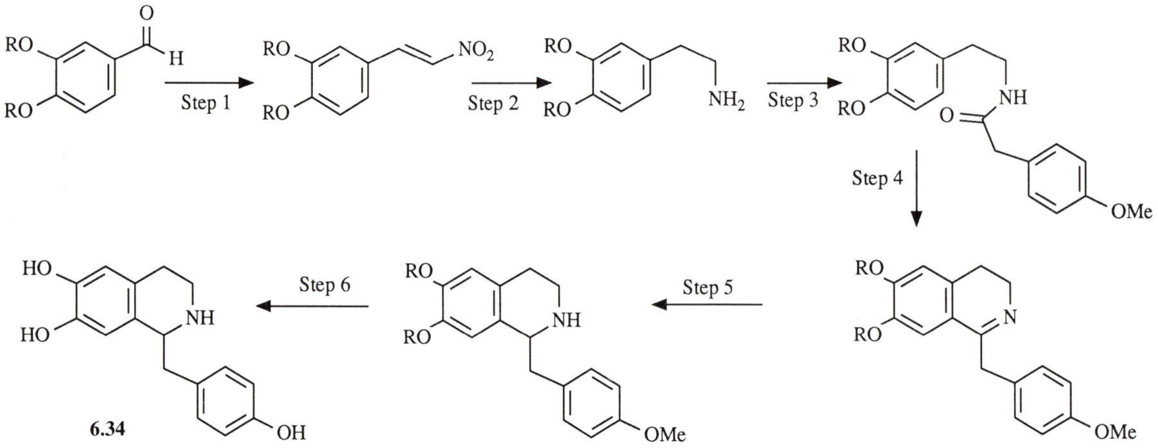

R = CH$_2$Ph

6.7 References

Adams, R. and Sloan, A.W. (1941). *Organic syntheses*, Coll. Vol. I, 478 (a real blood-and-thunder preparation of quinoline).

Claret, P.A. (1979). In *Heterocyclic chemistry* (ed. P.G. Sammes) (Vol. 4 of *Comprehensive organic chemistry*, ed. D. Barton and W.D. Ollis), p.155 (quinolines) and p.205 (isoquinolines). Pergamon Press, Oxford.

Furniss, B.S., Hannaford, A.J., Smith, P.W.G., and Tatchell, A.R. (1989). *Vogel 's textbook of practical organic chemistry* (5th edn), p.1185 (a rather more safety-conscious preparation of quinoline). Longman, Harlow.

Grethe, G. (ed.) (1981). In *Isoquinolines* (*The chemistry of heterocyclic compounds* [ed. A. Weissburger and E.C. Taylor], Vol. 3, Part 1). Wiley Interscience, New York.

Jones, G. (1977, 1990). In *Quinolines* (*The chemistry of heterocyclic compounds* [ed. A. Weissburger and E.C. Taylor], Vol. 32, Parts 1, 2, and 3). Wiley Interscience, New York.

Kathawala, G.F., Coppola, G.M., and Schuster, H.F. (ed.) (1989). In *Isoquinolines* (*The chemistry of heterocyclic compounds* [ed. A. Weissburger and E.C. Taylor], Vol. 3, Part 2). Wiley Interscience, New York.

Manske, R.H.F. and Kalka, M. (1953). *Organic reactions*, **7**, 59 (Skraup synthesis).

Whaley, W.M. and Govindachari, T.R. (1951). *Organic reactions*, **6**, p.151 (Pictet–Spengler synthesis).

7. Indoles

7.1 Introduction

Fusion of a benzene ring onto the C2/C3 positions of pyrrole formally produces the corresponding benzopyrrole **7.1** known as indole. An analogous theoretical transformation can be envisaged to form benzofuran **7.2** and benzothiophene **7.3**. This chapter will concentrate exclusively on indole, by far the most important member of this series.

Indole is a ten-π electron aromatic system. As with pyrrole, delocalisation of the lone pair of electrons from the nitrogen atom is necessary for aromaticity. The single overall electronic structure of indole is not completely described by structure **7.1**, because this implies localisation of the lone pair on the nitrogen atom. Mesomeric representation **7.1a** makes a contribution to the electronic structure of indole, as to a lesser extent do mesomeric representations where the negative charge occurs on the benzenoid ring.

A consequence of this delocalisation is that the lone pair is not available for protonation under moderately acidic conditions so, like pyrrole, indole is another weakly basic heterocycle. Another similarity to pyrrole is that being an 'electron-rich' heterocycle indole easily undergoes aromatic electrophilic substitution, and is also rather unstable to oxidative (electron-loss) conditions. However, an important difference emerges here, in that whereas pyrrole preferentially reacts with electrophiles at the C2/C5 positions, indole substitutes selectively at the C3 position. The reasons for this will be discussed later.

Neutrotransmitters are naturally-occurring substances which effect chemical communication between nerve cells by binding at specific sites on the cell surface called receptors.

Historically, interest in indoles arose with the isolation and characterisation of members of the enormous family of indole alkaloids, such as lysergic acid **7.4**. Many indole alkaloids possess interesting and sometimes useful biological activities. Although natural product chemistry is still an active area of primarily academic research, considerably more effort is expended nowadays in the preparation of indole derivatives as potential drug candidates. Following on from the observations that certain indole alkaloids or their semi-synthetic derivatives (e.g. lysergic acid diethylamide, LSD **7.5**) have potent central nervous system activity, it was established that the simple indole 5-hydroxytryptamine **7.6** is a major neurotransmitter. Many indole derivatives which mimic or block the binding of this neurotransmitter to its receptors have been synthesised and are beginning to find use in the treatment of various psychological disorders.

7.4 X = OH
7.5 X = NEt$_2$

7.6

7.2 Synthesis of indoles

As might be expected for a large branch of heterocyclic chemistry, many syntheses of indoles have been developed. We shall restrict our discussion to two, commencing with the widely-used Fischer synthesis.

The Fischer synthesis is the condensation of an aryl hydrazine with a ketone followed by cyclisation of the resultant hydrazone under acidic conditions to give the corresponding indole, as illustrated by the preparation of 2-phenyl indole **7.9**.

7.7   **7.8**   **7.9**

The actual cyclisation stage is not as imponderable as it appears. The first step is the acid-catalysed equilibration between hydrazone **7.8** and ene hydrazine **7.10**. The next step, which is irreversible, is a concerted electrocyclic reaction, forming a strong carbon–carbon bond, and breaking a weak nitrogen–nitrogen bond. The resulting imine **7.11** immediately re-aromatises by tautomerisation to aniline **7.12**. Finally, acid-catalysed elimination of ammonia forms indole **7.9**, reminiscent of the last step of the Knorr pyrrole synthesis (Chapter 2).

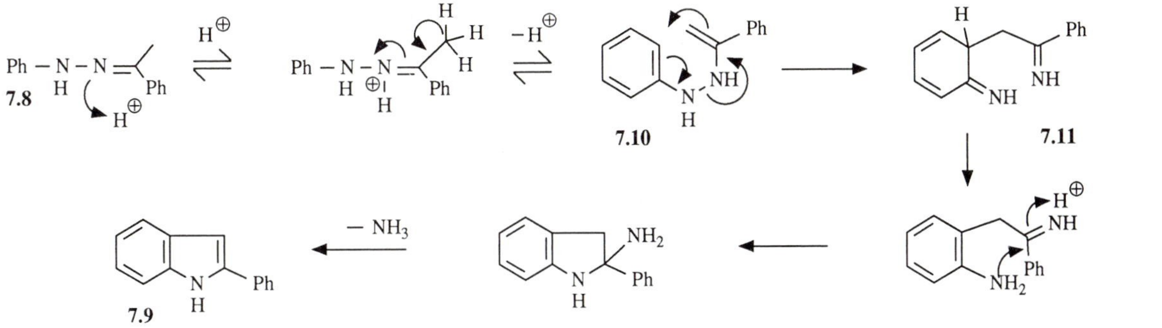

The electrocyclic reaction is very similar to the Claisen rearrangement of phenyl allyl ether **7.12** to give phenol **7.13**.

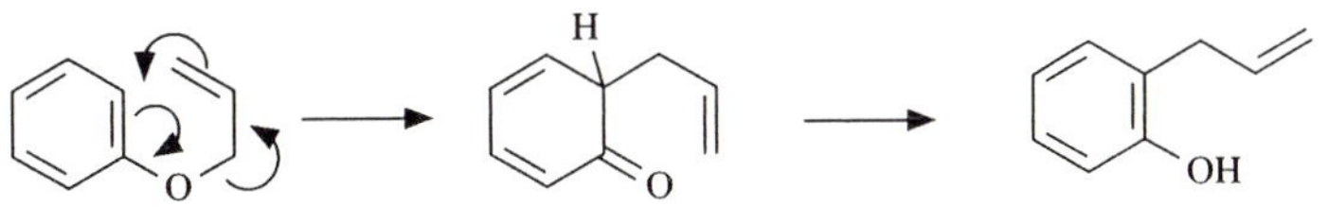

Some examples of the Fischer indole synthesis are shown below.

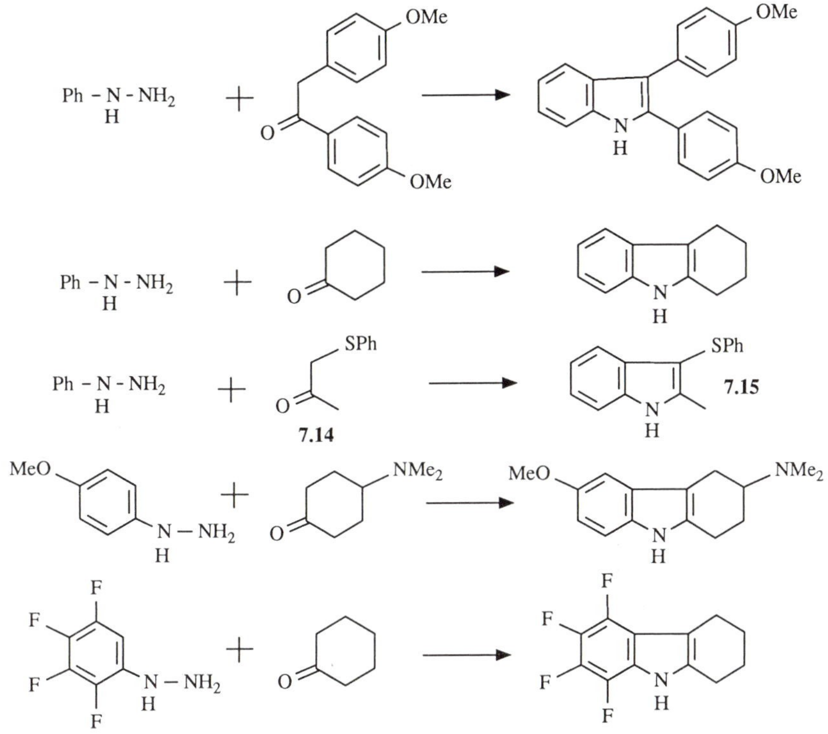

Cope rearrangement

Claisen rearrangement

Aza-Cope rearrangement

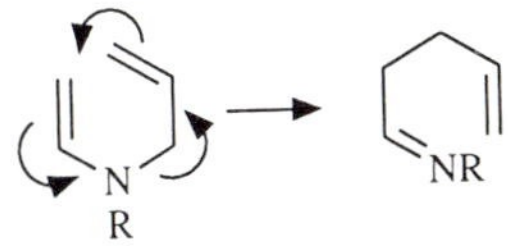

Diaza-Cope rearrangement

An interesting regioselectivity question arises with the use of unsymmetrical ketone **7.14** to prepare indole **7.15**. Two ene hydrazines **7.16**

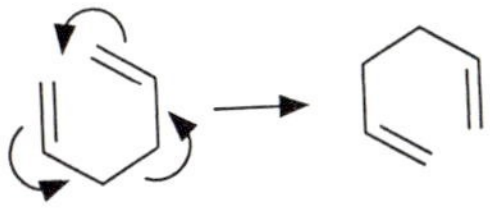

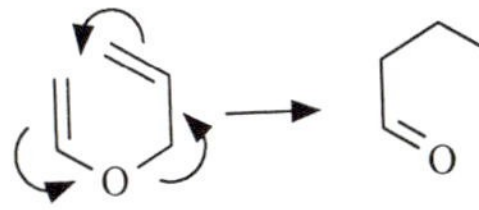

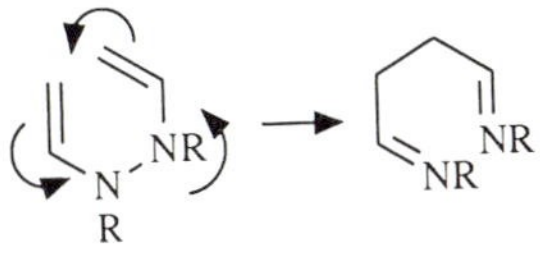

and **7.17** can form, which would give rise to indoles **7.15** and **7.18** respectively.

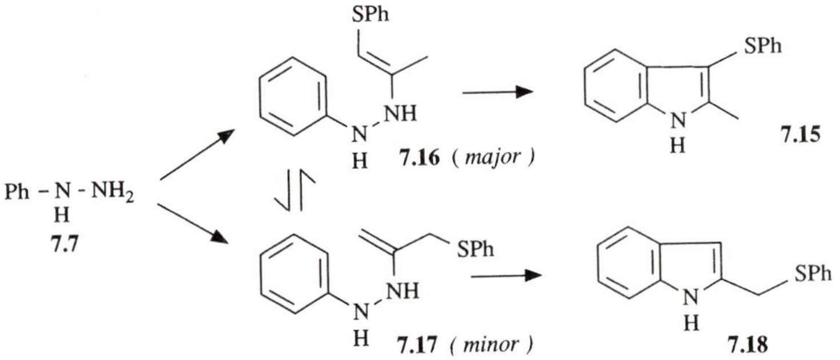

In such cases the most thermodynamically stable ene hydrazine, i.e. the one with the more highly substituted double bond, forms preferentially. In this particular example there is also extra stabilisation derived from conjugation of the lone pairs of electrons on the sulphur atom with the double bond. This regioselectivity in ene hydrazine formation is then reflected in the regioselectivity of indole formation.

The more recent Leimgruber synthesis is illustrated by the aminomethylenation of nitrotoluene **7.19** to give **7.20**, followed by hydrogenation to produce indole **7.1**.

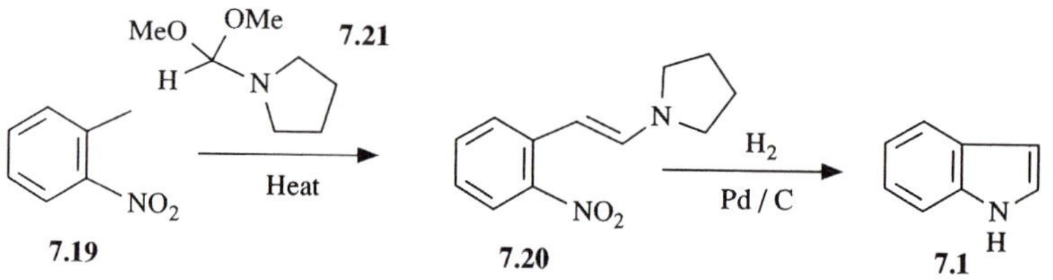

The combination of formyl pyrrolidine acetal **7.21** and nitrotoluene **7.19** produces electrophilic cation **7.22** and nucleophilic carbanion **7.23a,b** which react together affording enamine **7.20**.

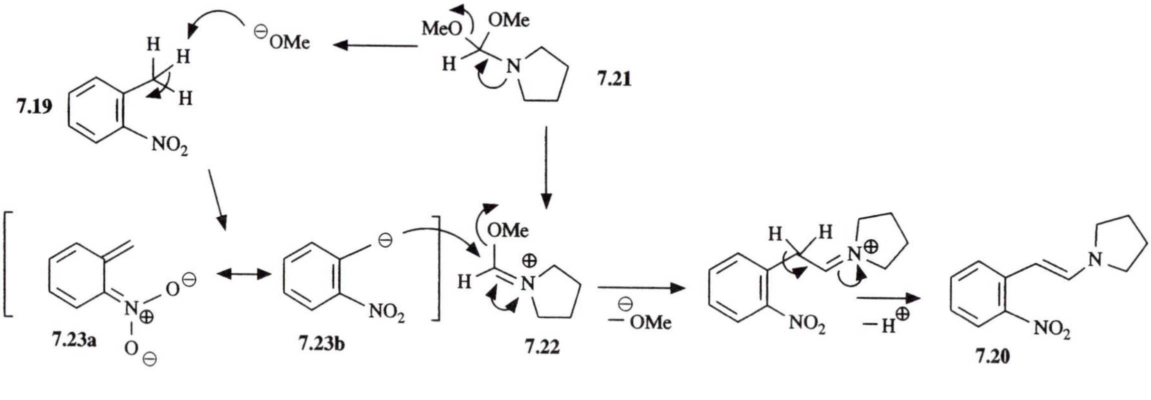

Hydrogenation of enamine **7.20** reduces the nitro group giving aniline **7.24**, then elimination of pyrrolidine produces indole **7.1**. Note the similarity of this ring closure step to the last step of the Fischer synthesis. In both cases the eventual C2 carbon atom is formally at the carbonyl oxidation level, even though it occurs as either an imine (Fischer synthesis) or an enamine (Leimgruber synthesis). Elimination of ammonia or pyrrolidine respectively is analogous to a condensation process involving elimination of water (as in the Knorr pyrrole synthesis).

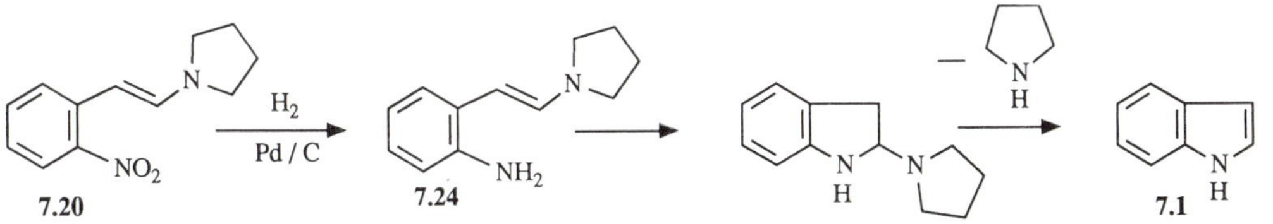

Some examples of the Leimgruber synthesis are shown below.

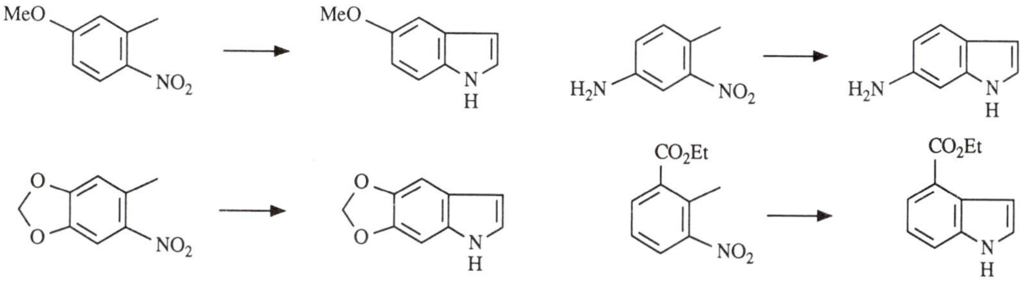

7.3 Electrophilic substitution of indoles

As an electron-rich heterocycle, indole easily undergoes electrophilic substitution. However whereas pyrrole reacts preferentially at the C2/C5 positions (see Chapter 2), indole reacts preferentially at the C3 position.

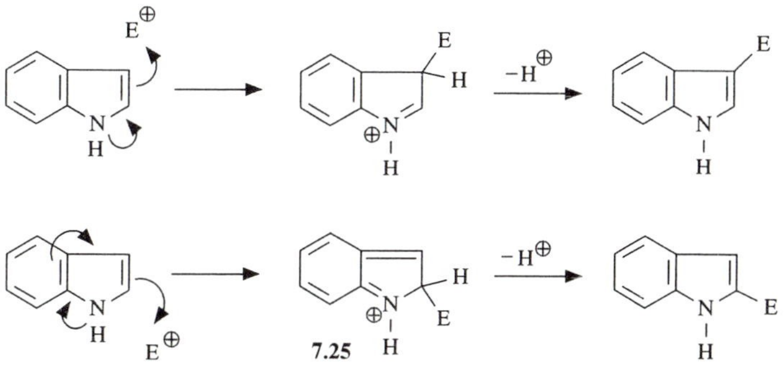

One explanation is that attack at C2 results in disruption of the aromaticity of the benzenoid ring, as in intermediate **7.25**. This is therefore a high-energy intermediate, and this reaction pathway is slower because the first step is rate-determining. Also the C3 selectivity is in accord with the electrophile attacking the site of highest electron density on the ring. In essence, indole tends to react like an enamine towards electrophiles, with substitution occurring at the C3 position, although substitution occurs at the C2 position when the C3 position is blocked.

Indole itself is unstable to the mineral acid conditions for nitration. The nitration of substituted indoles is quite complex and the outcome is dependent on the precise reaction conditions.

Like pyrrole, indole readily undergoes the Mannich reaction affording the aminomethyl derivative **7.26**. A variety of nucleophiles can displace the amine *via* an elimination followed by a 1,4-addition reaction, as in the preparation of acetate **7.27**.

This is the reactive electrophilic species of the Mannich reaction.

$CH_2 = \overset{\oplus}{N}Me_2$

The Vilsmeier reaction proceeds extremely well with indoles giving aldehydes such as **7.28**.

This is the reactive electrophilic species of the Vilsmeier reaction.

Aldehyde **7.28** is another useful synthetic intermediate, readily undergoing condensation reactions with active methylene compounds such as malonic acid and nitromethane to produce **7.29** and **7.30**.

Acylation of the C3 position can also be accomplished with acid chlorides, as illustrated in the synthesis of indole **7.34**, a drug for the treatment of depression. Reaction of indole **7.31** with oxalyl chloride affords C3-substituted product **7.32** even though the benzene ring is very electron-rich. Conversion to amide **7.33** is followed by reduction with lithium aluminium hydride which removes both carbonyl groups, affording the target indole **7.34**.

7.4 Anion chemistry of indole

Treatment of indole ($pK_a = 17$) with strong bases such as butyl lithium, Grignard reagents, or metal hydrides produces the corresponding indolyl anion, which reacts with electrophiles either on nitrogen or at the C3 position. With lithium, sodium, or potassium as counterion the indolyl anion tends to react on nitrogen, as in the preparation of **7.35**. However, with magnesium as the counterion the intermediate has an essentially covalent rather than ionic structure, and reaction tends to occur at the C3 position, as in the preparation of **7.36**.

When the nitrogen is blocked, deprotonation can occur at the C2 position, adjacent to the electronegative heteroatom. This offers a means of introducing electrophiles at this position, complementing the C3 selectivity shown by classical electrophilic substitution. For instance, alcohol **7.37** can be prepared in this way using ethylene oxide as the electrophile.

7.5 Problems

1. Devise a synthesis of the antidepressant drug **7.38**.

7.38

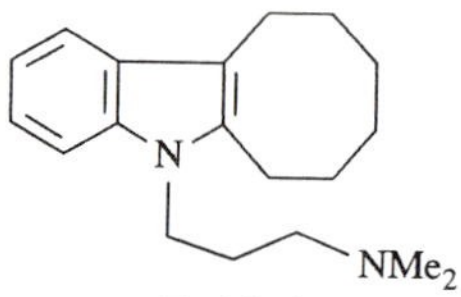

2. The synthesis of amino ester **7.41** is shown below. What is the mechanism of the conversion of **7.39** to **7.40**.

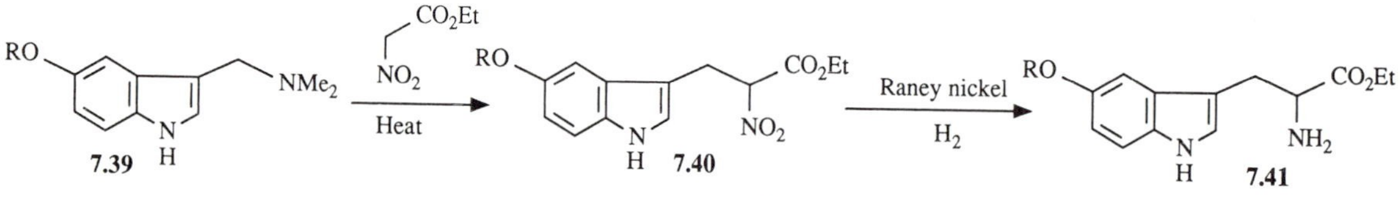

7.39   **7.40**   **7.41**

R = PhCH$_2$

3. It was intended to prepare imine **7.43** from indole **7.42** by deprotonation at the C2 position then quenching with benzonitrile followed by an aqueous workup. However, the isolated products were ketone **7.44** and sulphonamide **7.45**. Account for this observation.

7.42
1. *n* - BuLi
2. Ph—C≡N
3. HCl / H$_2$O

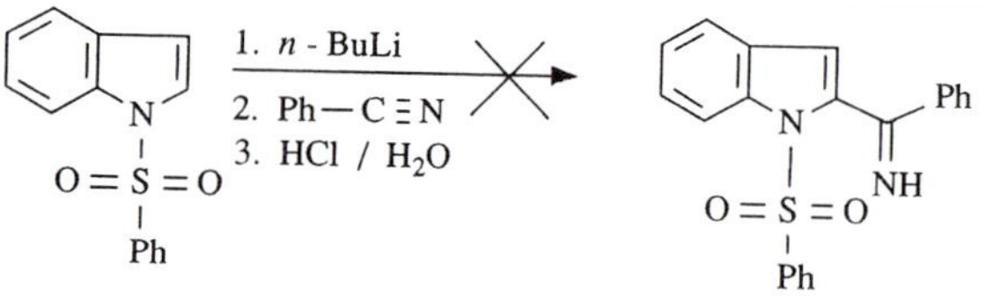

7.43

7.44   **7.45**

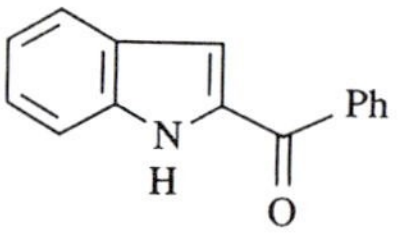

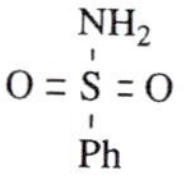

7.6 References

Brown, R.T. and Joule, J.A. (1979). In *Heterocyclic chemistry* (ed. P.G. Sammes) (Vol. 4 of *Comprehensive organic chemistry*, ed. D. Barton and W.D. Ollis), p.411 (indoles and related systems). Pergamon Press, Oxford.

Furniss, B.S., Hannaford, A.J., Smith, P.W.G., and Tatchell, A.R. (1989). *Vogel's textbook of practical organic chemistry* (5th edn), p.1161 (preparation of indole **7.9**). Longman, Harlow.

Houlihan, W.J. (ed.) (1972). *Indoles* (The chemistry of heterocyclic compounds [ed. A. Weissburger and E.C.Taylor], Vol. 25, Parts 1 – 3). Wiley Interscience, New York.

Leimgruber, W. (1985). *Organic syntheses*, **63**, 214 (indole synthesis).

Robinson, B. (1969). *Chem. rev.*, **69**, 227 (Fischer indole synthesis).

Saxton, J.E. (ed.) (1979). *Indoles* (The chemistry of heterocyclic compounds
[ed. A. Weissburger and E.C. Taylor], Vol. 25, Part 4). Wiley Interscience, New York.

Sundberg, R.J. (1970). *The chemistry of indoles*. Academic Press, New York.

8. Five-membered ring heterocycles with three or four heteroatoms

8.1 Introduction

The broad category of five-membered ring heterocycles containing three or four heteroatoms encompasses many heterocyclic systems. Obviously there is considerable variation in the physical and chemical properties of such a large group of heterocycles. For instance, with regard to aromaticity, oxadiazole **8.3** is considered to be less aromatic than triazole **8.8** or tetrazole **8.9**.

Note the parallel with furan being less aromatic than pyrrole, Chapter 2.

8.1   **8.2**   **8.3** oxadiazoles   **8.7**   triazoles **8.8**

8.4   **8.5**   **8.6** thiadiazoles

8.9 tetrazole   **8.10** oxatriazole   **8.11** thiatriazole

Nevertheless, this collection of heterocycles does share certain characteristics. The trend we have seen of decreasing tendency towards electrophilic substitution on going from furan, pyrrole, and thiophene to the azoles is continued into these series. The presence of additional 'pyridine-like' nitrogen atoms renders these systems particularly 'electron-deficient', and electrophilic substitution is of little importance.

Conversely, nucleophilic substitution (which we have seen in earlier chapters on 1,3-azoles and pyridines) does occur in these systems, especially when the carbon atom concerned is between two heteroatoms, as in the displacement reactions of oxadiazole **8.12** and tetrazole **8.13**.

Once again note the analogy with
standard carbonyl chemistry.

Another similarity with azoles is that there are examples of deprotonation
of alkyl substituents between two heteroatoms followed by quenching the
resultant carbanions with electrophiles, as in the preparation of oxadiazole
8.14.

Ring deprotonation is also known with certain members of these series.
Carbanion **8.15** is stable at low temperature (-70°C) and can be trapped with
electrophiles, but on warming to room temperature it decomposes with ring
fragmentation and extrusion of nitrogen. This fragmentation process is
reminiscent of the base-catalysed cleavage of isoxazoles (Chapter 4).

For simplicity we shall now consider the synthesis of just three members
of these series, 1,2,4-oxadiazole **8.3**, 1,2,3-triazole **8.7**, and tetrazole **8.9**.

8.2 Synthesis of 1,2,4-oxadiazoles

Disconnection of the C5-oxygen bond in **8.16** leads to iminoalcohol **8.17**
which occurs as amide **8.18**. Cleavage of the amide linkage leads to an
activated carboxylic acid **8.20** plus the heteroatom-containing amidoxime
8.19.

8.16   **8.17**   **8.18**   **8.19**   **8.20**

X = Leaving group

Amidoximes can be prepared by acid-catalysed additon of hydroxylamine to nitriles.

An example of this approach to oxadiazoles is shown by the conversion of ester **8.21** to oxadiazole **8.22**, prepared as a potential candidate for the treatment of senile dementia. Simple esters are metabolically unstable in man because of the high activity of esterases. These enzymes catalyse the hydrolysis of esters to carboxylic acids. A common tactic in drug research when confronted with the problem of metabolic instability of a biologically active ester is to replace the ester group with a small heterocycle (often oxadiazole), to try to produce a biologically-active molecule with improved metabolic stability. This concept of replacing fragments of a molecule by groups with broadly similar physicochemical parameters in a systematic manner is known as bioisosteric replacement. In this instance oxadiazole **8.22** can mimic both the physical and biological properties of **8.21**, but it is obviously not a substrate for esterases.

8.21   **8.22**

8.3 Synthesis of 1,2,3-triazoles

These are best prepared by a 1,3-dipolar cycloaddition of an azide and an acetylene.

For instance, triazole **8.8** itself has been prepared by cycloaddition of hydrazoic acid to acetylene.

Although a simple mechanism can be drawn for this transformation, it is only useful as a 'book-keeping exercise' to ensure that the correct structure is drawn for the product. In reality the reaction is a concerted process and the usual considerations of nucleophilic and electrophilic attack do not apply.

Excellent yields are achieved in these cycloadditions when electron-withdrawing groups are present on the acetylene, as in the preparation of triazole **8.23**.

8.4 Synthesis of tetrazoles

Tetrazole itself explodes on heating with loss of two molecules of N_2.

Tetrazoles of general structure **8.24** can be prepared in a very similar manner to triazoles, except that nitriles are used rather than acetylenes. Once again the reaction with azides is a concerted cycloaddition process.

Let us now consider the synthesis of tetrazole **8.27**, an inhibitor of the enzyme ornithine decarboxylase, which catalyses the conversion of ornithine **8.25** to diamine **8.26**.

The tetrazole moiety is an excellent bioisosteric replacement for a carboxylic acid, being a small, polar, acidic heterocycle.

Tetrazole **8.27** is sufficiently similar to ornithine **8.25** in its physical properties to bind to the active site of the enzyme. However, as it obviously cannot undergo the decarboxylation process, it acts as an inhibitor of the enzyme.

The synthesis commences with alkylation of the stabilised carbanion derived from cyanoester **8.29** with iodide **8.28** to give adduct **8.30**

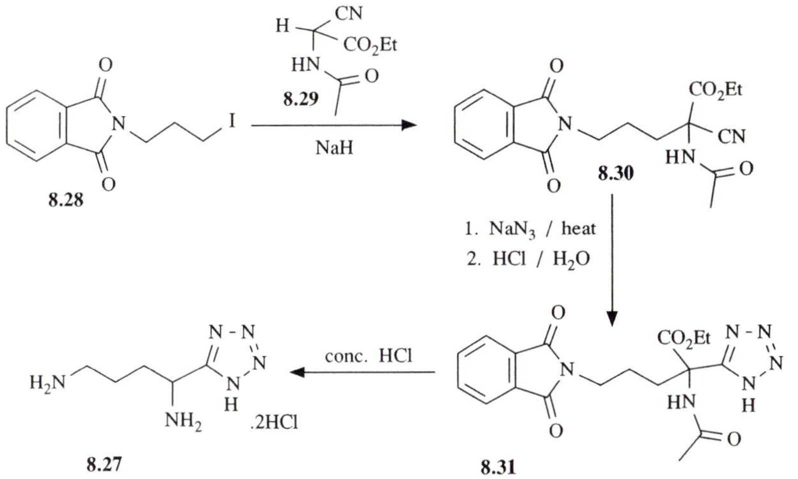

Cycloaddition with sodium azide followed by acidification during aqueous workup affords tetrazole **8.31**.

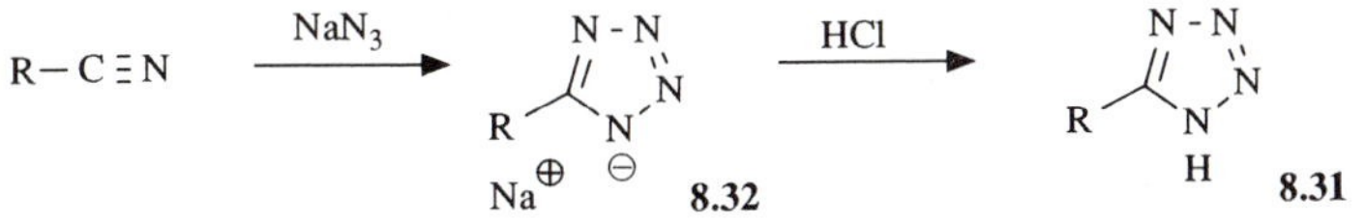

Note that the first-formed product from the cycloaddition is actually the sodium tetrazolate salt **8.32**. Protonation affords the neutral tetrazole **8.31**. Prolonged acidic hydrolysis accomplishes several transformations: hydrolytic removal of both the phthalimide and acetyl nitrogen protecting groups, and hydrolysis/decarboxylation of the ester. The net result is to produce the target tetrazole **8.27** as its dihydrochloride salt. This tetrazole-assisted decarboxylation is mechanistically very similar to the decarboxylation of malonyl half-esters **8.33**.

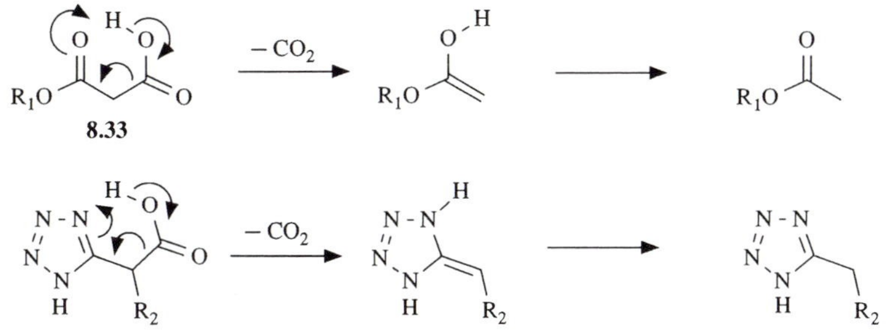

8.5 Problems

1. Triazoles and tetrazoles can be alkylated on nitrogen under basic conditions, as in the synthesis of the clinically-used antifungal drug **8.35** in which 1,2,4-triazole is alkylated by a chloromethyl ketone and an epoxide, both good alkylating agents. What is the mechanism of formation of epoxide **8.34**? Of compounds **8.34** and **8.35**, which is achiral and which is racemic?

2. What is the mechanism of formation of oxadiazole **8.22**?

8.6 References

Butler, R.N. (1977). *Adv. heterocyclic chem.*, **21**, 323 (tetrazoles).

Clapp, L.B. (1976). *Adv. heterocyclic chem*, **20**, 65, (1,2,4-oxadiazoles) .

Gilchrist, T.L. (1985). *Heterocyclic chemistry*, p.81 (1,3-dipolar cycloadditions in heterocyclic synthesis). Longman, Harlow.

Gilchrist, T.L. and Gymer, G.E. (1974). *Adv. heterocyclic chem.*, **16**, 33 (1,2,3-triazoles).

Grimmett, M.R. (1979). In *Heterocyclic chemistry* (ed. P.G. Sammes) (Vol. 4 of *Comprehensive organic chemistry*, ed. D. Barton and W.D. Ollis), p.357 (triazoles and tetrazoles). Pergamon Press, Oxford.

9. Six-membered ring heterocycles containing one oxygen atom

9.1 Introduction

The pyrilium cation **9.1**, 2-pyrone **9.2**, 4-pyrone **9.3**, and their benzo-fused analogues the benzopyrilium cation **9.4**, coumarin **9.5**, chromone **9.6**, are the parent structures of a series of six-membered ring heterocycles containing one oxygen atom. The impetus for research in this area comes from the enormous number of plant-derived natural products based on the benzopyrilium, coumarin, and chromone structures.

The red, violet, and blue pigments of flower petals are called anthocyanins, and are glycosides of various benzopyrilium cations. Delphinidin chloride **9.7**, for example, is a blue pigment. Khellin **9.8** is a natural product which has found clinical application in the treatment of bronchial asthma and has been the starting point for the design of many totally synthetic chromones with improved biological properties.

In natural product chemistry, the acetal formed between an aliphatic or aromatic alcohol and a sugar is termed a glycoside.

Coumarin **9.5** is itself a natural product which occurs in lavender oil and has been found in over sixty species of plants.

The pyrilium cation **9.1** is the oxygen analogue of pyridine and is a six π-electron aromatic system. Nevertheless, being a cation it is reactive towards nucleophiles and is readily hydrolysed to give dialdehyde **9.9**. These reactions are reversible, a fact which has been used in a synthesis of **9.1** from **9.9**. At low pH (high acidity) the equilibrium lies to the side of the pyrilium species **9.1** but if the medium is basified then hydrolysis of **9.1** occurs to give **9.9**. This is because one mole of hydroxide is consumed on going from pyrilium cation **9.1** to neutral aldehyde **9.9**. Increasing the hydroxide concentration therefore forces the equilibrium from left to right.

The carbonyl groups of 4-pyrone and 4-pyridone absorb at approximately 1650 cm^{-1} and 1550 cm^{-1} respectively. The lower energy of the pyridone absorption reflects greater single bond character, and hence greater delocalisation.

In contrast, 2- and 4-pyrones are considered to have relatively little aromatic character. Whereas in an analogous nitrogen series 4-pyridone **5.23** has significant aromatic character (mesomeric representation **5.23a** making a considerable contribution to the overall electronic distribution), aromatic mesomeric representation **9.3a** makes less of a contribution to the overall electronic structure of 4-pyrone. As with furan, the higher electronegativity of oxygen leads to heterocycles of little aromaticity in cases where delocalisation of electron density from the heteroatom is a prerequisite for that aromaticity.

Let us now consider the synthesis of a pyrilium salt, a coumarin, and a chromone.

9.2 Synthesis of a pyrilium salt

A typical pyrilium salt synthesis is illustrated by the preparation of salt **9.12**. The precursor to **9.12** is pyran **9.11**, available by dehydration of 1,5-diketone **9.10**. Note the similarity of this sequence to the Hantzch pyridine synthesis, Chapter 5. Also, the dehydrative cyclisation of a diketone to oxygen heterocycle **9.11** is reminiscent of furan synthesis, Chapter 2.

One hydrogen atom has to be removed from the C4 position of pyran **9.11** to produce the pyrilium cation, but it is important to realise that the hydrogen atom is lost not as a proton but as a *negatively-charged hydride ion*. The process is therefore an *oxidation* of pyran **9.11**.

9.13 **9.14a** **9.14b** **9.15**

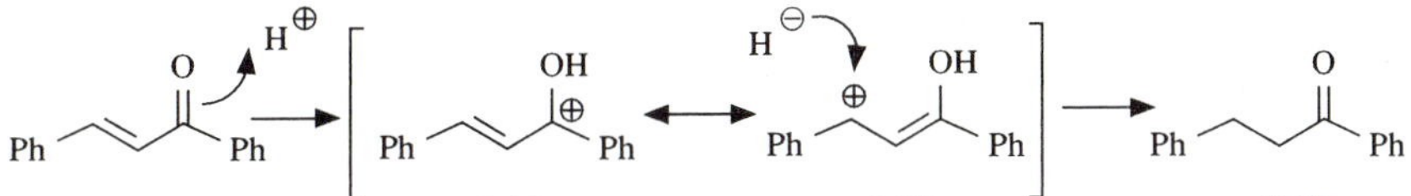

A suitable oxidant is cation **9.14a,b**, derived from α,β-unsaturated ketone **9.13** by protonation under strongly acidic conditions in the absence of water. Quenching of this cation with a hydride ion (from the C4 position of **9.11**) produces the saturated ketone **9.15**. The balanced equation is shown below.

9.13 **9.11** **9.12** **9.15**

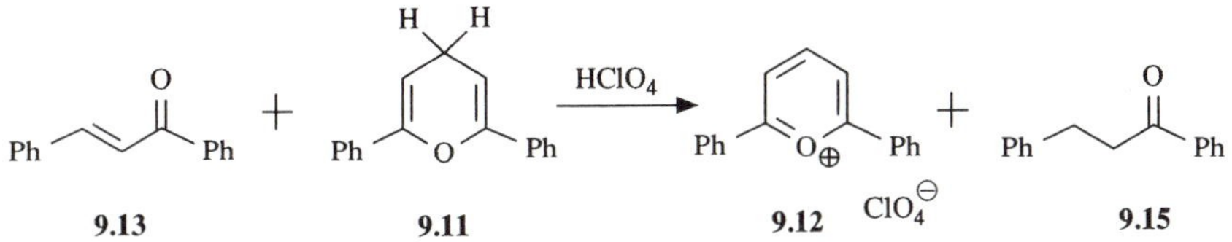

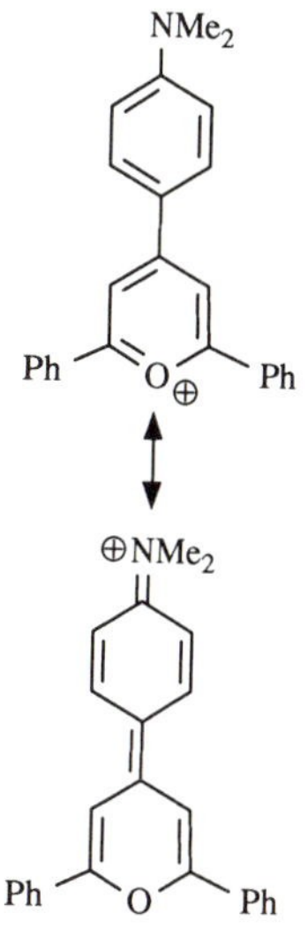

Most pyrilium salts have electron-donating aromatic substituents at the C2, C4, or C6 positions which serve to stabilise the positive charge by resonance.

9.3 Synthesis of coumarins

Let us consider the synthesis of bromocoumarin **9.16**, a compound which exhibits biological activity against parasitic trematodes that cause schistosomiasis, a very common disease in the tropics. Retrosynthetic cleavage of lactone **9.16** gives diester **9.17**, which in principle can be derived from condensation of *ortho*-hydroxybenzaldehyde **9.18** and diethyl malonate.

9.16 **9.17** **9.18**

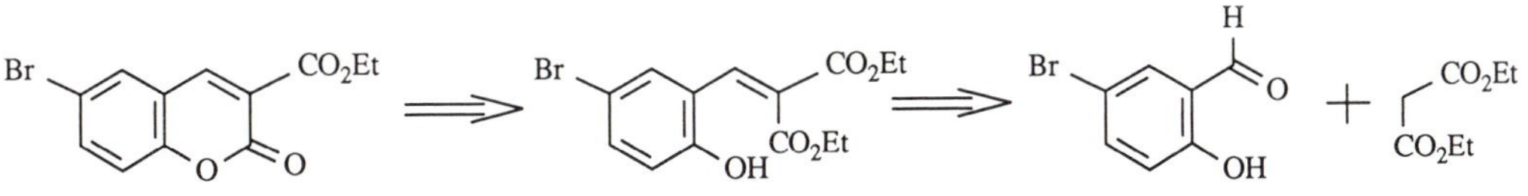

In practise a Knoevenagel condensation reaction yields coumarin **9.16** directly, without isolation of diester **9.17**. The mechanism is shown below .

9.16 **9.17**

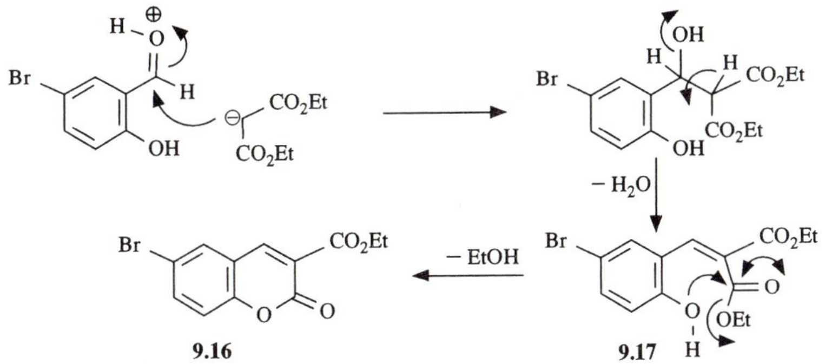

9.4 Synthesis of chromones

Let us consider the synthesis of flavone **9.19**, which is the parent of a large series of natural products. Disconnection of the carbon–oxygen bond in the usual way results in enol **9.20** which exists as 1,3-diketone **9.21**. This 1,3-dicarbonyl relationship can be exploited in the classical manner yielding *ortho*-hydroxyacetophenone **9.22**. The synthetic problem centres on methodology for the C-benzoylation of the enolate derived from **9.22** with some activated benzoic acid derivative **9.23**.

| **9.19** | **9.20** | **9.21** | **9.22** | X = Leaving group |

In practice, the Konstanecki–Robinson synthesis of chromones commences with O-benzoylation not C-benzoylation, to afford ester **9.24**. Base-catalysed rearrangement produces the required 1,3-diketone **9.21**, *via* intramolecular benzoylation of the intermediate enolate. Acid-catalysed dehydration then affords flavone **9.19**.

9.5 Reactions with nucleophiles

Although some examples of electrophilic substitution are known, the chemistry of these series is dominated by nucleophilic ring-opening reactions, sometimes followed by ring-closure to give new heterocycles. For instance, aminolysis of **9.1**, **9.2**, and **9.3** leads to pyridine **5.1** and pyridones **5.22** and **5.23**.

The mechanism of the conversion of 4-pyrone to 4-pyridone involves an initial Michael reaction followed by ring-opening. Tautomerisation of enol **9.25** to aldehyde **9.26**, followed by cyclisation, affords 4-pyridone **5.23**.

The reaction of pyrilium salts with nucleophiles may involve electrocyclic ring-opening of the intermediate dienes, as in the formation of ketone **9.27**.

A similar susceptibility to nucleophilic attack is observed in the benzo-fused series. Coumarin **9.5** is hydrolysed by hydroxide to carboxylate salt **9.28**. This process is reversible, and acidification regenerates the lactone.

An important difference between the monocyclic and benzo-fused series is that reactions with amines do not lead to the corresponding heterocycles in the benzo-fused series. For instance, aminolysis of chromone **9.29** affords phenol **9.30**. Benzopyridone **9.32** is not produced. The facile tautomerisation between **9.25** and **9.26** would analogously give ketone **9.31** in this series. This high-energy intermediate is not aromatic, and the reaction stops at phenol **9.30**.

Phenols do not exist or react in their tautomeric keto forms.

9.6 Problems

1. What is the mechanism of the conversion of pyrone **9.2** to pyridone **5.22** by aminolysis?

2. Explain the formation of pyrazole **9.33**.

3. How can chromone **9.34** be converted to **9.35**?

4. What is the mechanism of this cyclisation?

9.7 References

Horing, E.C. *et al.* (1955). *Organic synthesis*, Coll. Vol. III, 165 (experimental details of a Knoevenagel condensation to give a coumarin ester).

Livingstone, R. (1977). In Rodd's *Chemistry of carbon compounds*, Vol. IV, p.2 (pyrilium salts; 2- and 4-pyrones); p.69 (benzopyrilium salts); p.96 (coumarins); p.138 (chromones). Elsevier, Amsterdam.

Staunton, J. (1979). In *Heterocyclic chemistry* (ed. P.G. Sammes) (Vol. 4 of *Comprehensive organic chemistry*, ed. D. Barton and W.D. Ollis), p.607 (pyrilium salts); p.629 (2-pyrones and coumarins); p.659 (4-pyrones and chromones). Pergamon Press, Oxford.

Wheeler, T.S. (1963). *Organic synthesis*, Coll. Vol. IV, 479 (experimental details for the preparation of flavone).

10. Pyrimidines

10.1 Introduction

Formal replacement of a CH unit in pyridine **5.1** by a nitrogen atom leads to the series of three possible diazines, pyridazine **10.1**, pyrimidine **10.2**, and pyrazine **10.3**. Like pyridine they are fully aromatic heterocycles. The effect of an additional nitrogen atom as compared to pyridine accentuates the essential features of pyridine chemistry. Electrophilic substitution is difficult in simple unactivated diazines because of both extensive protonation under strongly acidic conditions and the inherent lack of reactivity of the free base. Nucleophilic displacements are comparatively easier.

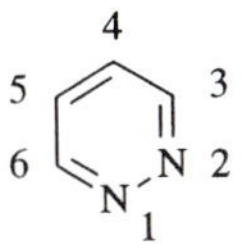

10.1 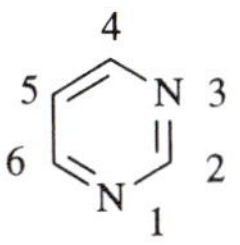   **10.2** 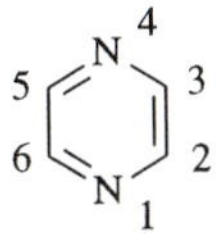   **10.3** 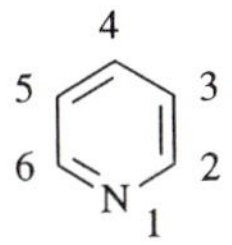   **5.1**

Interestingly, the second electronegative heteroatom reduces the capacity of the diazines to tolerate the positive charge resulting from protonation. Pyridazine **10.1** (pK_a= 2.24), pyrimidine **10.2** (pK_a = 1.23), and pyrazine **10.3** (pK_a = 0.51) are all far less basic than pyridine (pK_a = 5.23).

The most important of the diazines is pyrimidine **10.2**. Pyrimidine derivatives uracil **10.4**, thymidine **10.5**, and cytosine **10.6** are the monocyclic 'bases' of nucleic acids. The bicyclic bases are the purines adenine **10.7** and guanine **10.8**. The purine ring is essentially a fusion of the pyrimidine and imidazole rings.

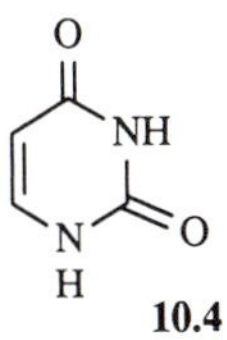

10.4 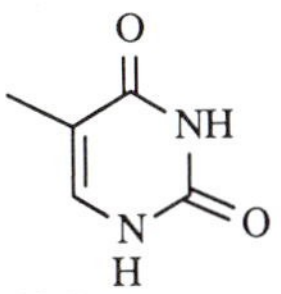   **10.5** 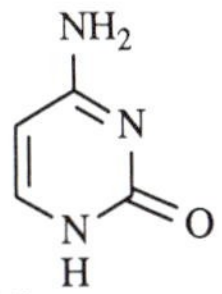   **10.6** 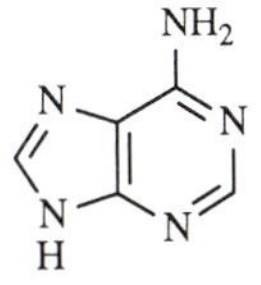   **10.7** 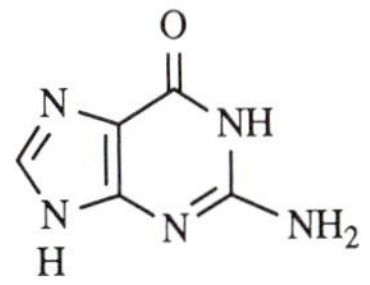   **10.8**

The actual biosynthesis of purines (illustrated below in abbreviated form for the nucleotide adenosine monophosphate AMP **10.9**) involves construction of a pyrimidine ring onto a pre-formed imidazole.

Nucleotides are the monomeric building blocks of deoxyribonucleic acid (DNA) in which is stored the genetic information of the cell.

The enzymes that manipulate nucleotides, nucleic acids, etc. are the points of therapeutic intervention for a number of diseases involving cell replication disorders such as cancers and viral infections. For instance, AZT **10.10**, an inhibitor of the enzyme reverse transcriptase, is an anti-viral drug currently used in the treatment of AIDS.

We shall now go on to consider the synthesis and chemistry of the pyrimidine ring system.

10.2 Synthesis of pyrimidines

Disconnection of the N1–C6 bond in generalised pyrimidine **10.11** in the usual way produces enol **10.12**, which exists as ketone **10.13**. Similarly, disconnection of the carbon–nitrogen double bond in **10.13** yields a dicarbonyl compound **10.14** and an amidine **10.15**. This retrosynthetic analysis, suggesting the combination of bis-electrophilic and bis-nucleophilic components, is the basis of a very general pyrimidine synthesis.

Where R_4 is a hydrogen or carbon atom, **10.15** is simply an amidine. However, urea **10.16**, thiourea **10.17**, or guanidine **10.18** and their derivatives may be used. These nucleophiles may be condensed with ester and nitrile functionalities as well as with aldehydes and ketones. Such condensations to afford pyridimidine derivatives are usually facilitated by acid or base catalysis, although certain combinations of reactive electrophilic and nucleophilic compounds require no catalyst at all. Some examples are shown below.

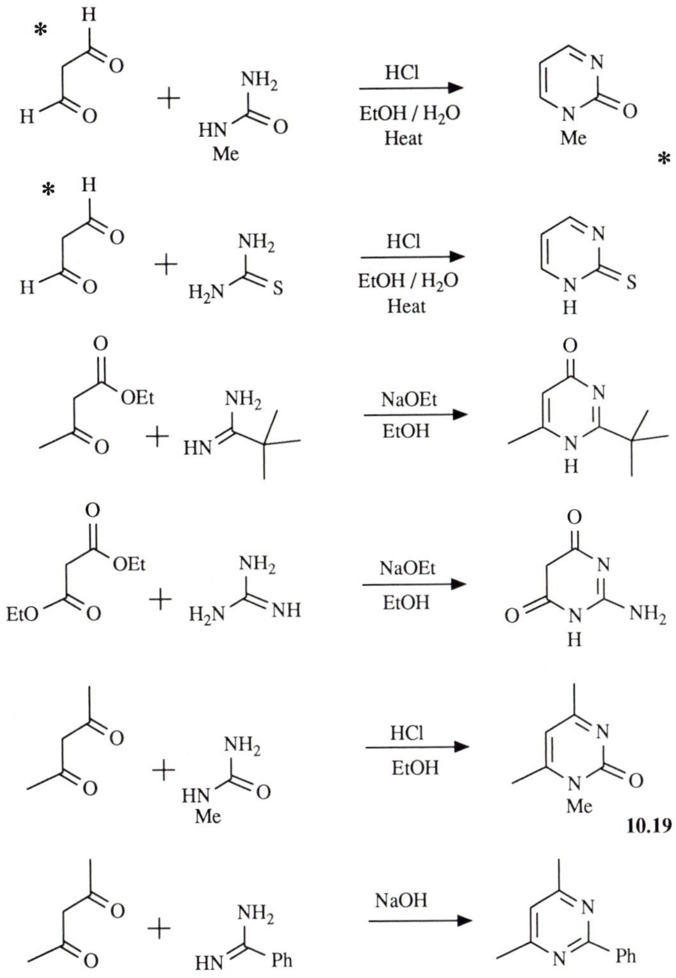

* Prepared by *in situ* hydrolysis of

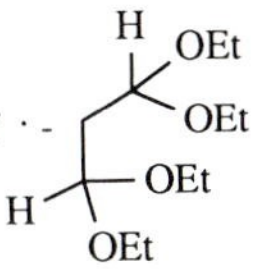

Note that several of these examples produce pyrimidones, analogous to the pyridones previously encountered in Chapter 5. A representative mechanism is shown for the preparation of 2-pyrimidone **10.19**, and is simply two consecutive condensations.

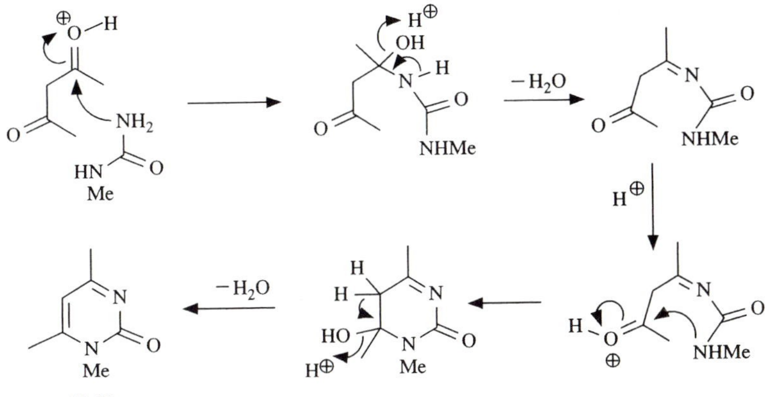

10.19

10.3 Electrophilic substitution of pyrimidones

As mentioned earlier, electrophilic substitution on unactivated pyrimidines is of little importance. But, as with pyridine, the pyrimidine nucleus can be activated towards electrophilic attack by employing N-oxides or pyrimidones, for the same reasons as were discussed in Chapter 5.

For instance, nitration of 2-pyrimidone **10.20** affords nitropyrimidone **10.21**. With doubly-activated systems such as **10.22**, nitration to give **10.23** can occur without heating.

10.4 Nucleophilic substitution of pyrimidines

Leaving groups at the C2, C4, and C6 positions of pyrimidines can be displaced by nucleophiles, with the negative charge of the intermediate delocalised over both nitrogen atoms.

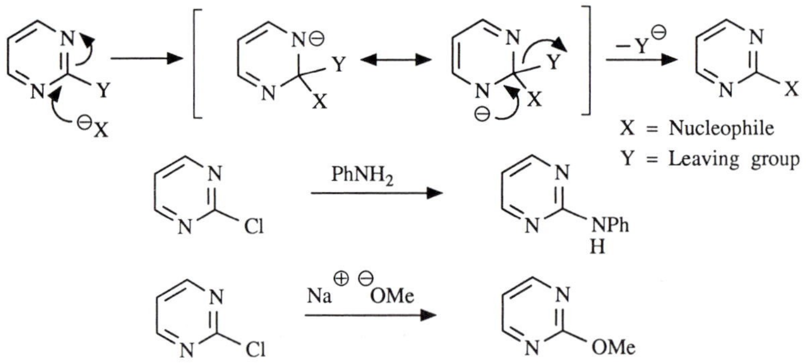

X = Nucleophile
Y = Leaving group

Chlorinated pyrimidines themselves are often accessible from the corresponding pyrimidones by reaction with phosphorous oxychloride. (Again, see Chapter 5 for an explanation of this sort of reaction.) For instance, aminopyrimidine **10.24** can be synthesised by the classical sequence depicted below.

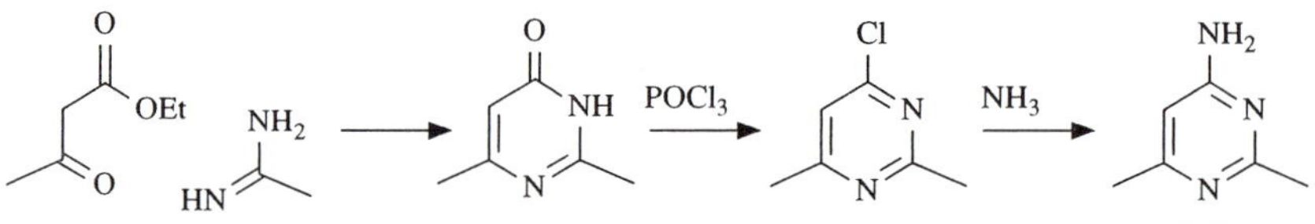

10.5 Problems

1. Write a mechanism for this nitration, but starting from an alternative mesomeric representation of **10.20** that helps to explain the increased susceptibility of such pyrimidones to electrophilic attack.

10.20   **10.21**

2. Barbiturates (pyrimidine triones such as **10.25**) used to be widely used as sedatives, but have now largely been superseded by drugs with fewer side-effects. Suggest a synthesis of **10.25**.

10.25

3. There are several preparations of cytosine **10.6** available, one of which is the condensation of nitrile **10.26** with urea **10.16**. Propose a mechanism for this reaction.

10.26   **10.16**   **10.6**

10.6 References

Brown, D.J. (1962). In *The pyrimidines (The chemistry of heterocyclic compounds* [ed. A. Weissburger and E.C. Taylor], Vol. 16). Wiley Interscience, New York.

Brown, D.J. (1970). In *The pyrimidines (The chemistry of heterocyclic compounds* (ed. A. Weissburger and E.C. Taylor], Vol. 16, Supplements 1 and 2). Wiley Interscience, New York.

Furniss, B.S., Hannaford, A.J., Smith, P.W.G., and Tatchell, A.R. (1989). *Vogel's textbook of practical organic chemistry* (5th edn), p.1177 (preparation of barbiturate **10.25**). Longman, Harlow.

Hurst, D.T. (1980). *An introduction to the chemistry and biochemistry of pyrimidines, purines, and pteridines.* Wiley, New York.

11. Answers to problems

11.1 Answers to problems in Chapter 2

Note that the reaction proceeds with attack of the amino group on the least hindered ketone.

1. Reaction of diketone **2.46** with aminoketone **2.47** produces enamine **2.48** which is not isolated, but cyclises directly to give pyrrole **2.43**.

2. The lone pair of electrons of **2.44** is delocalised on to the carbonyl group as shown, increasing the electron density at the aldehydic carbon atom. This renders it less reactive to nucleophilic attack.

Under acidic conditions alcohol **2.45** readily gives cation **2.49a,b** which is stabilised by a similar delocalisation of the nitrogen lone pair.

This highly electrophilic species then reacts with alcohol **2.45** to give dimer **2.50**. Repetition of this process leads to polymeric material.

3. As discussed in Chapter 2, interception of cation **2.31** with a nucleophilic counterion such as acetate produces the 2,5-addition product **2.32**. Tetrafluoroborate is a non-nucleophilic counterion and hence the only pathway available to **2.31** is loss of a proton to give nitrofuran **2.33** directly.

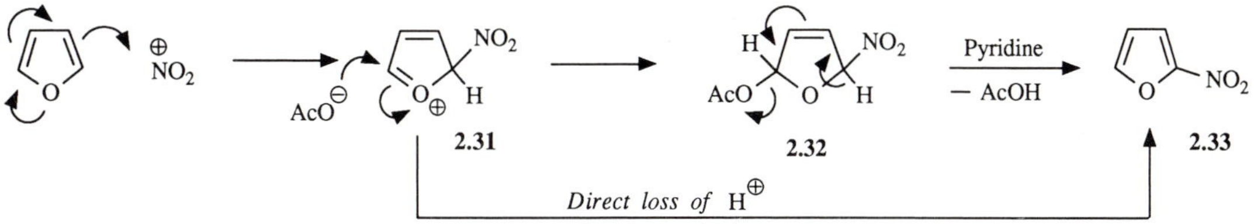

4. The mechanism is a straightforward Friedel–Crafts acylation.

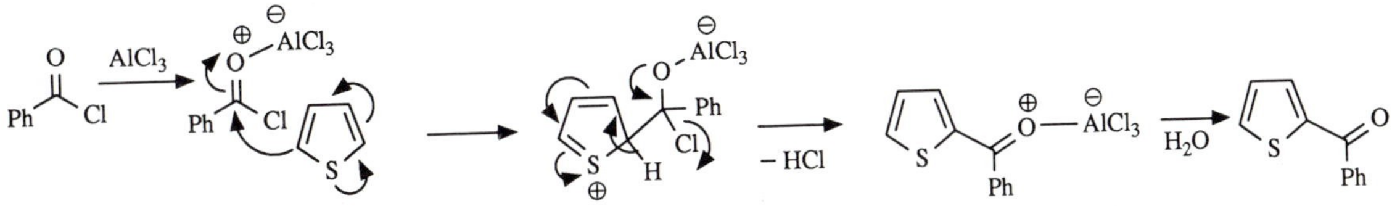

11.2 Answers to problems in Chapter 3

1. Application of our generalised oxazole retrosynthesis leads to a simple glycine derivative.

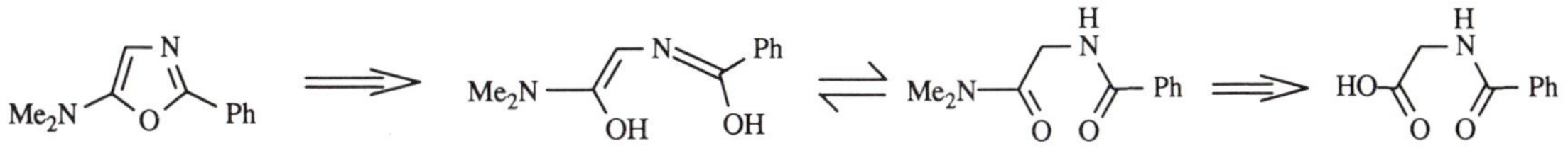

The forward synthesis is shown below:

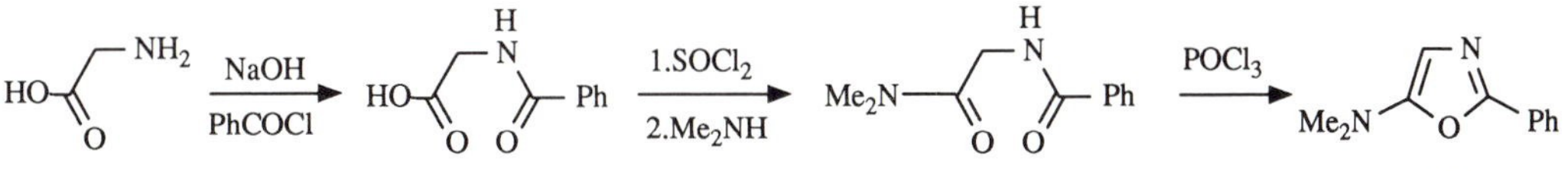

2. The mechanism of this oxazole formation is identical to that of the Hantzch thiazole synthesis. However, because of the reduced nucleophilicity of a carbonyl group as compared to a thiocarbonyl (due to the higher electronegativity of oxygen), this synthesis only proceeds under vigorous conditions (high temperatures, amide component as solvent, etc).

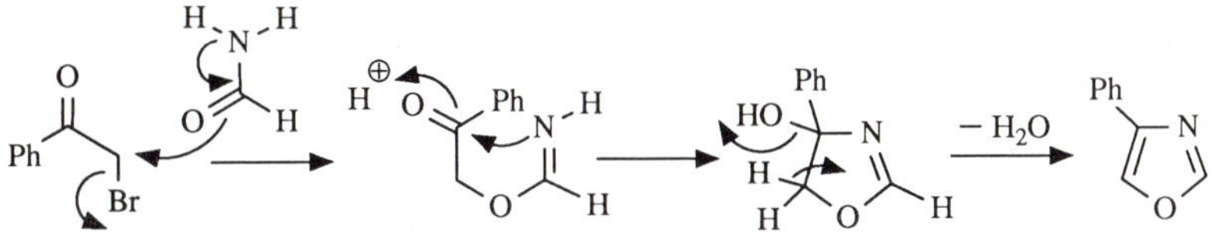

The alternative sequence would give a positional isomer of oxazole **3.44**.

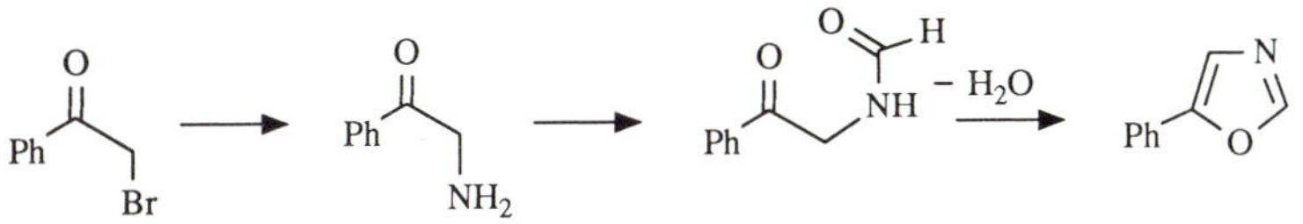

3. Bromination of **3.45** gives a bromoketone which is condensed with thiourea to give aminothiazole ester **3.47**. This is then hydrolysed to acid **3.46**.

Ketone **3.45** itself is readily prepared by nitrosation of ethyl acetoacetate followed by O-methylation.

11.3 Answers to problems in Chapter 4

1. The reaction is a Michael addition followed by elimination of cyanide ion.

2. The overall strategy is to protect the nitrogen of pyrazole (as an acetal), deprotonate, introduce the side chain as an electrophile, then deprotect.

3. Oxidation of oxime **4.41** produces nitrile oxide **4.46** which cyclises to isoxazole **4.47**.

4. Reaction with hydroxylamine occurs on the aldehyde group of the more reactive minor tautomer **4.43** affording isoxazole **4.44**. Methoxide-induced fragmentation as shown gives enolate **4.48** which is quenched by a proton in the workup to afford 2-cyanocyclohexanone **4.45**.

4.43 ... **4.44** ... **4.45** ... **4.48**

11.4 Answers to problems in Chapter 5

1. The process is essentially analogous to a Michael reaction.

2. A reasonable mechanism is:-

3. Pyridyl amide **5.39** is easily metallated at the C3 position. Quenching with the aldehyde, and cyclisation of the resulting alcohol **5.42** onto the amide group, produces lactone **5.40**.

4. The reaction probably proceeds *via* enaminoester formation then cyclisation.

In fact nucleophilic substitution of pyridine N-oxides occurs more easily than on simple pyridines, as the nitrogen atom is positively charged.

11.5 Answers to problems in Chapter 6

1. The quinolone synthesis involves an addition–elimination reaction followed by an intramolecular aromatic acylation.

The displacement reaction occurs by initial nucleophilic attack on the benzenoid ring (with the negative charge being delocalised onto the oxygen atom as shown) then elimination of chloride ion. The presence of the fluorine substituent is essential for this displacement, activating the ring towards nucleophilic attack by its electron-withdrawing inductive effect.

2. *Step 1*. Condensation of the aldehyde with nitro methane under basic conditions produces the α,β-unsaturated nitro compound.

Step 2. Lithium aluminium hydride was used, although hydrogenation can also effect this type of reduction.

Step 3. Acylation of the amine with an acid chloride in the presence of an appropriate base gave the amide.

Step 4. This isoquinoline formation is of course an example of the Bischler–Napieralski synthesis, although phosphorous trichloride was actually used in this example, not phosphorus oxychloride.

Step 5. Sodium borohydride was used to reduce the imine to the amine.

Step 6. The catecholic and phenolic ethers were removed by treatment with hydrobromic acid. Benzyl ethers are frequently removed by reduction (e.g. hydrogenation) but reduction, of course, would not remove the methyl ether. The mechanism of the deprotection is shown below.

11.6 Answers to problems in Chapter 7

1. Indole **7.38** was prepared by a Fischer indole synthesis followed by N-alkylation as shown.

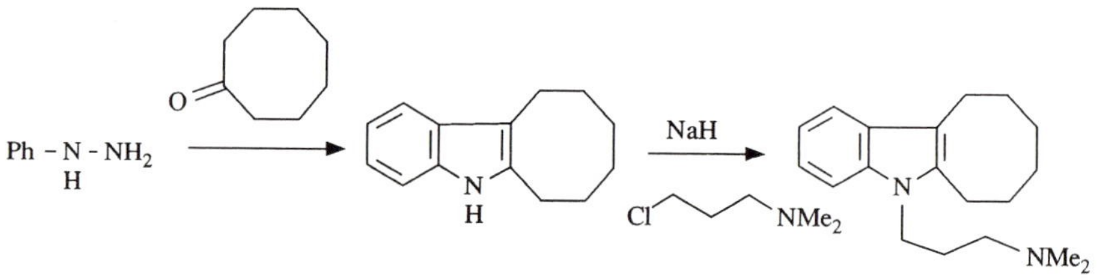

7.38

2. The nitroacetate (pK_a = 6.79) protonates the tertiary amine functionality of indole **7.39**, facilitating the elimination of methylamine to give cation **7.46**. Conjugate addition of the nitroacetate anion then produces **7.40**.

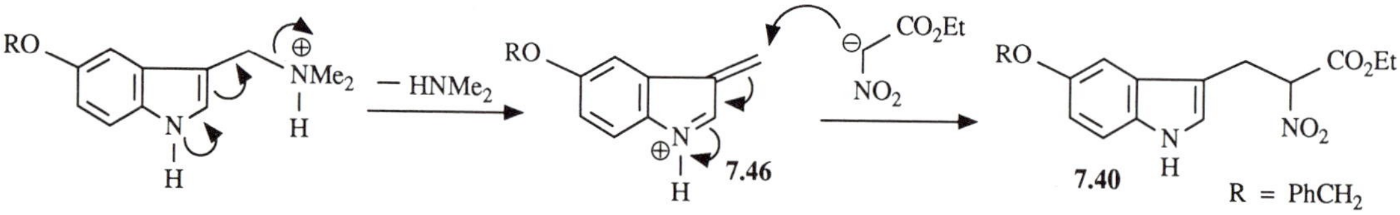

R = PhCH$_2$

3. As intended, C2 carbanion **7.47** attacked the nitrile giving **7.48**, which unexpectedly attacked the adjacent sulphonyl group giving indolyl anion **7.49**. During the acidic aqueous workup this anion is quenched and the reactive N-sulphonyl imine functionality is readily hydrolysed affording ketone **7.44** and sulphonamide **7.45**.

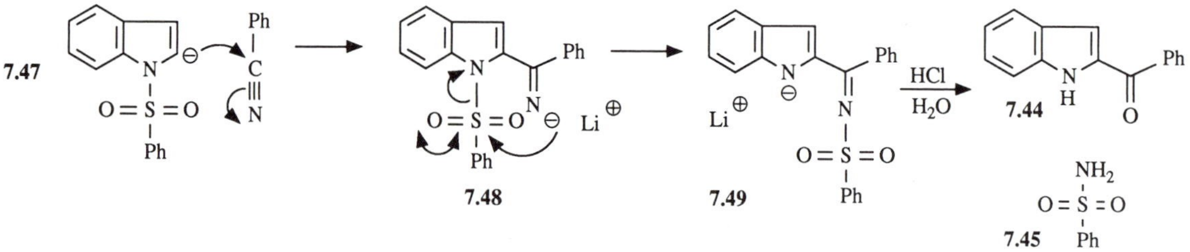

11.7 Answers to problems in Chapter 8

1. Sulphur ylid **8.36** is the key intermediate in the formation of epoxide **8.34**. Epoxide **8.34** is racemic but alcohol **8.35** is achiral.

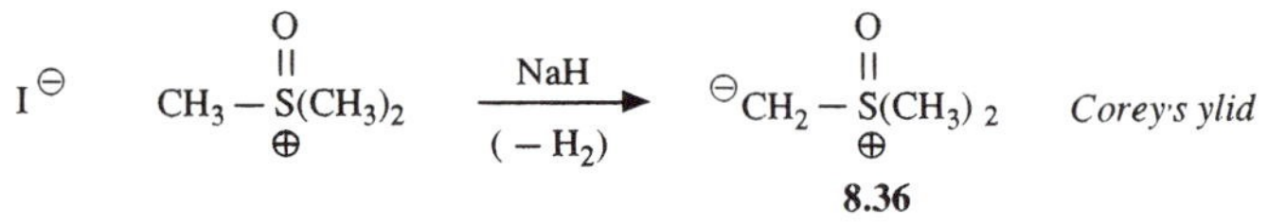

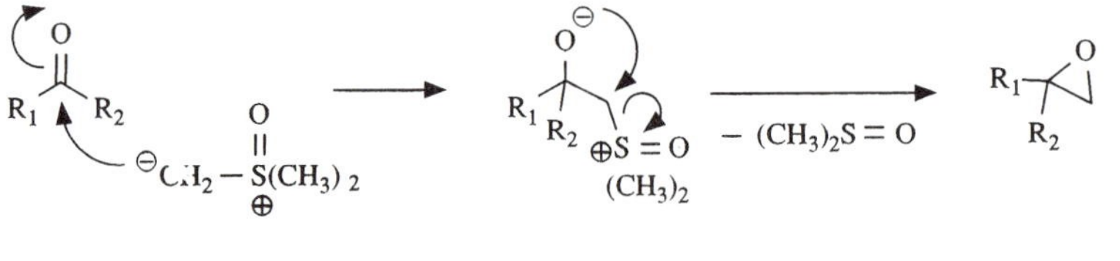

2. This oxadiazole formation involves O-acylation of the amidoxime followed by a condensation.

11.8 Answers to Problems in Chapter 9

1. The mechanism is similar to the 4-pyrone example.

2. The first stage is the same as the preparation of **9.30**, then cyclisation affords the pyrazole.

3. This is a Mannich reaction (see Chapter 2) and is an unusual example of an electrophilic substitution on a chromone.

4. The reaction is a straightforward acid-catalysed condensation, passing through carbonium ion **9.36a,b**.

11.9 Answers to problems in Chapter 10

1. The overall electronic distribution of 2-pyrimidone has a considerable contribution from mesomer **10.20a**.

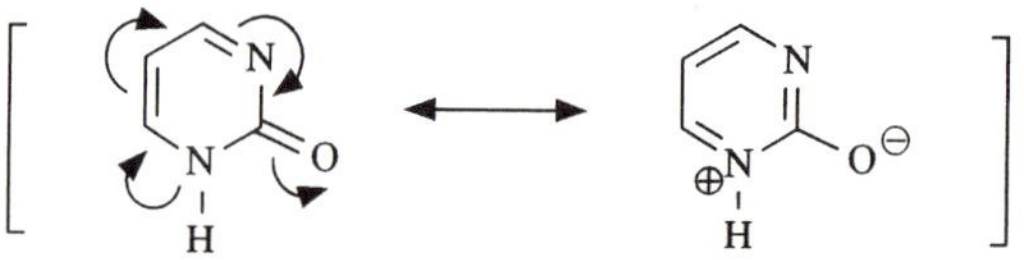

10.20 **10.20a**

The mechanism of nitration is shown below.

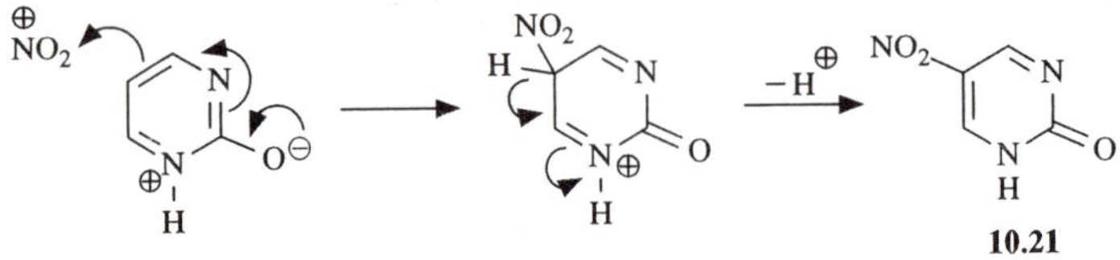

10.21

2. Disconnection of barbiturate **10.25** produces bis-electrophile **10.27** and urea. In practice malonate ester **10.28** (X = OEt) is used.

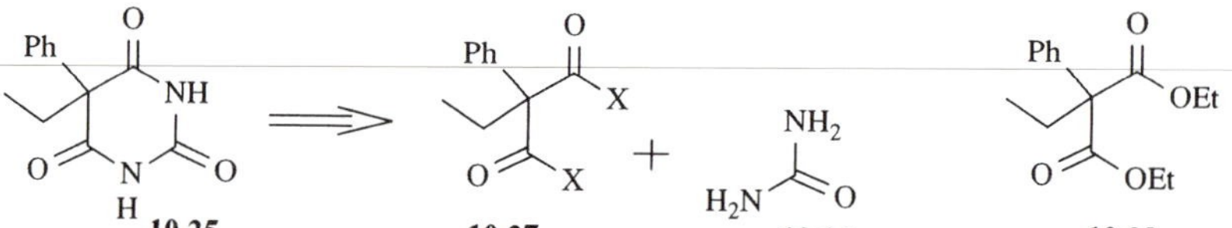

10.25 **10.27** **10.16** **10.28**

3. Hydrolysis of acetal **10.26** leads to reactive aldehyde **10.29** *in situ*.

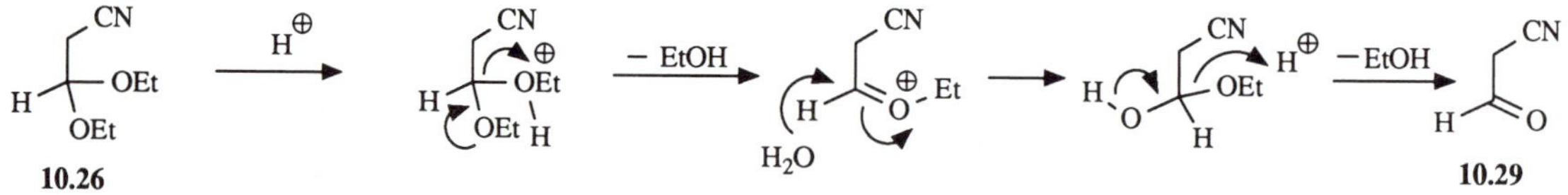

10.26 **10.29**

Condensation of aldehyde **10.29** with urea followed by cyclisation onto the nitrile produces cytosine **10.6**. Observe how cyclisation onto a nitrile affords the amino functionality directly, as compared with the three step sequence used in the synthesis of **10.24** where an ester is used in the cyclisation step.

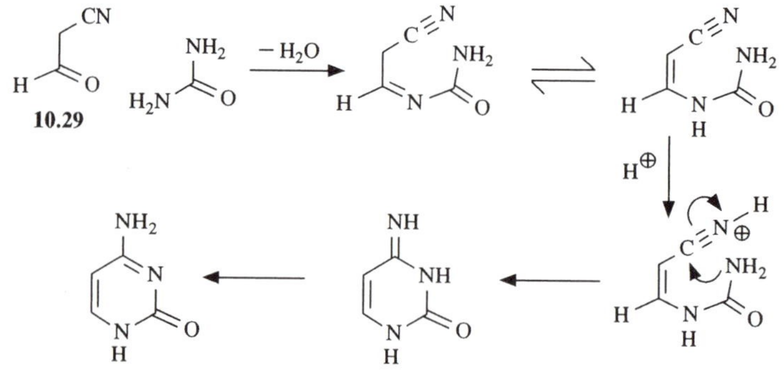

10.29

10.6

Index

α-effect 29
acetyl nitrate 15
acid chlorides 6
α-aminoketone 13
ammonia 5
AMP, biosynthesis of 74
anion chemistry of
  furan 17
  imidazole 25
  indole 59
  isoquinoline 50
  isothiazole 32
  isoxazole 32
  pyrazole 32
  pyridine 42
  pyrrole 17
  quinoline 50
  thiazole 25
anthocyanins 67
aryl hydrazines 54
1,2 azoles 28
1,3 azoles 20
AZT 74

barbiturates 77
benzene 2
benzopyrilium cation 67
bioisosteric replacement 63
Bischler–Napieralski synthesis
    (isoquioline) 48

Chichibabin reaction 40
chlorophyll 11
chromone 67
Claison rearrangement 55
condensation 3
Cope rearrangement 55
coumarin 67
cytosine 73

delocalisation 3
delphinidin chloride 67
dihydropyridines 37
disconnection 4
DNA 73
drugs for the treatment of
  AIDS 74
  asthma 31
  bacterial infection 27
  depression 60
  fungal infection 41, 66
  inflammation 22
  schizophrenia 18
  senile dementia 45, 63
  sleep disorders 77
  trematode infection 69
  ulcers 1, 11

electrophilic substitution of
  furan 14
  imadazole 24
  indole 57
  isoquinoline 49
  isothiazole 32
  isoxazole 32
  oxazole 24
  pyrazole 32
  pyridine 37
  pyridine N-oxide 38
  pyridones 39
  pyrimidones 76
  pyrrole 14
  quinoline 49
  thiazole 24
  thiophene 14

Fischer synthesis (indole) 54
flavone 70
furan 10

Gould–Jacobson synthesis
(quinolone) 51

Hantzsch pyridine synthesis 36
Hantzsch thiazole synthesis 23
heteroaromaticity 1
histamine 20
hydrazine 29
hydrogen sulphide 12
hydroxylamine 29
5-hydroxytryptamine 54

imidazole 20
imidoyl halide 22
indole 53
isoquinoline 46
isothiazole 28
isoxazole 28

Khellin 67
Knorr synthesis (pyrrole) 14
Konstanecki–Robinson synthesis
    (chromone) 70

Leimgruber synthesis (indole) 56
lysergic acid 54

Mannich reaction 16, 58

neurotransmitters 54
nitrile oxides 30
nucleic acids 73
nucleophilic substitution of
    imidazoles 26
  isoquinolines 49

oxazoles 26
pyridines 40
pyrimidines 76
quinolines 49
thiazoles 26

ornithine 64
ortho-activating substituents 43
oxadiazoles 61
oxazoles 20
oxazolidinones 8

Paal–Knorr synthesis (pyrrole,
    thiophene, furan) 12
phosphorous oxychloride 15, 47
phosphorous sulphide 12
Pictet–Spengler synthesis
    (isoquinoline) 48
pyrazole 28
purine 73
pyridine 35
pyridine N-oxide 38
pyridine sulphur trioxide complex 15,
  35
pyridones 39
pyridontriazine 41
pyrrilium cation 67
pyrimidines 73
pyrones 67
pyrrole 10

Quinoline 46

resonance 2
retrosynthesis 4
Robinson–Gabriel synthesis
    (oxazole) 21

Skraup synthesis (quinoline) 46
synthesis of
  chromones 70
  coumarins 69
  furans 11
  heterocycles, principles of 3–8
  imidazoles 22
  indoles 54
  isoquinolines 46
  isothiazoles 29
  oxadiazoles 62
  oxazoles 21
  pyrazoles 29
  pyridines 35
  pyrilium salt 68
  pyrimidines 74
  pyrroles 11
  quinolines 46
  tetrazoles 64

thiazoles 23
thiophenes 11
triazoles 63

tetrazoles 61
thiamin 20

thiazoles 20
thiophene 10
thymidine 73
triazole 61

uracil 73

Vilsmeier formylation 15, 58